INDIAN RIVER CO. MAIN LIBRARY

3 2901 000 2468

P9-BYY-359

Indian River County Main Library
1600 21st Street
Vero Beach, FL 32960

THE CRYOTRON FILES

THE
CRYOTRON
FILES

The Untold Story of Dudley Buck, Cold War Computer Scientist and Microchip Pioneer

IAIN DEY & DOUGLAS BUCK

WITH RESEARCH BY ALAN DEWEY

THE OVERLOOK PRESS
NEW YORK, NY

This edition first published in hardcover in the United States in 2018 by
The Overlook Press, Peter Mayer Publishers, Inc.

141 Wooster Street
New York, NY 10012
www.overlookpress.com
For bulk and special sales, please contact sales@overlookny.com
or write to us at the above address.

Copyright © 2018 by Iain Dey and Douglas Buck

All rights reserved. No part of this publication may be reproduced or
transmitted in any form or by any means, electronic or mechanical, including
photocopy, recording, or any information storage and retrieval system now
known or to be invented, without permission in writing from the publisher,
except by a reviewer who wishes to quote brief passages in connection with
a review written for inclusion in a magazine, newspaper, or broadcast.

Cataloging-in-Publication Data is available from the Library of Congress

Book design and typeformatting by Bernard Schleifer
Manufactured in the United States of America

ISBN 978-1-4683-1577-6
First Edition
1 3 5 7 9 10 8 6 4 2

3 2901 00648 2468

To my mother, who was left to raise three small children as a widow. To Bobbie. To my brother David, who, together, kept digging for information.

—Douglas Buck

To Carla, for her endless support. To Allegra, for always making me smile. To Anne & Charley, for everything. And to Cristina, for always nudging me to put my hand up.

—Iain Dey

CONTENTS

THE CRYOTRON FILES

1
PROJECT LIGHTNING

RAINCLOUDS WERE HANGING OVER IDLEWILD INTERNATIONAL AIRport as the KLM flight from Amsterdam touched down on the tarmac. Sergey Lebedev peered out of the window, unimpressed. They had told him in Moscow that New York was at its best in April—bright and sunny, yet without the oppressive heat and humidity of summer. He had left his raincoat at home and advised the six Soviet computer experts joining him for the trip not to bother bringing theirs either.

The propellers of the Lockheed Super Constellation were still winding down as the group got to the top of the plane's steps. One by one they looked up at the gathering storm and realized their wardrobe error. Above the noise of the engines, Lebedev could hear the grumbling begin.

It was Sunday, April 19, 1959. They had left Moscow two days earlier, and all were in need of sleep.

Each man carried a black leather briefcase. Some contained drawings and notes about the biggest and best computers in the Soviet Union—information they planned to present to the Americans. Others were carrying vodka and black caviar to treat their hosts during what was scheduled to be a two-week tour of the United States.

They had come for a rapprochement. The US government had agreed to let the Russians see inside America's most secret computer labs; the Kremlin would offer the same courtesy in exchange.

Lebedev, at age fifty-six the Soviet Union's top computer expert, had been tasked with leading the delegation himself. During World War II he had built a system to stabilize the sights of tank cannons.

He then created the first computer in the Eastern bloc with a small group of researchers at the University of Kiev, which in turn led to him being handpicked by Joseph Stalin to lead the USSR's computer effort. He had retained the role under the new premier, Nikita Khrushchev, and was finally starting to make progress with his inventions.

Although the Soviet Union had caught up with America on the nuclear bomb and had beaten the Americans into space with the launch of Sputnik some fifteen months earlier, computer technology was one area where the Americans had a sizable advantage.

Stalin was an obstacle to the development of Soviet computer technology. He had objected to the development of any machine that would replicate the human brain or replace a man on a factory production line; he saw it as a capitalist evil. That had forced Lebedev and his contemporaries to develop computers with very strictly defined military missions: for translation, weather forecasting, and to calculate the firing range of missiles.

America, on the other hand, had burned billions of dollars on a sprawling mass of computer projects with undefined or moving objectives. Private companies were competing with universities and government departments for lucrative defense contracts to build computers for the army, the air force, the navy, or the newly created commercial honeypot that was NASA, the American space agency. It was a creative hotbed that had spawned a booming industry, one that was inventing ever more advanced technologies at breakneck speed.

Lebedev had built an impressive machine in his lab in Moscow, but had not worked out how to mass-produce the device effectively. The Americans, meanwhile, were already rolling out reliable computers by the hundreds.

American businesses were installing giant machines sold by the likes of IBM and RCA which could be used to run their payrolls or settle their taxes. Programs were under way to computerize air traffic control and US census data.

Both superpowers knew that computer technology had the power to change the dynamics of the Cold War. There were clear economic benefits to be gained from the digitization of the American economy. Yet there were also more direct military uses for computing power.

Both sides were developing nuclear missiles at great pace, and computers were needed to guide those missiles and to identify and shoot down any incoming enemy threats. The American science community was bubbling with stories about one young scientist in particular.

Dudley Buck at MIT had developed an ultrafast computer with no moving parts that would fit in a man's shirt pocket. Given that the most advanced computers at that time occupied whole floors of office buildings, it was an attention-grabbing concept. Buck had been touring America to educate academics and business leaders about his work. Although the term had not yet been coined, he had invented a prototype microchip named the Cryotron.

The Soviet Union was years behind on this technology, and that posed a serious problem for Lebedev. According to an article Lebedev had seen in *Life* magazine two years earlier, Buck's tiny computer chip would be used as the guidance system for America's new intercontinental ballistic missile. At the time the article was published, Buck's prototype device was a long way from being capable of deployment with a nuclear warhead. In the intervening period, however, a number of large research projects under the auspices of the US government had been set to drive forward Buck's Cryotron technology. Yet it was still not quite perfected.

The US State Department had given Lebedev and his team permission to see inside Buck's lab. Just three days' after Lebedev and his team of scientists touched down in New York, they were scheduled to meet Buck—and to see his invention for themselves.

DUDLEY BUCK WAS working late in his lab yet again. Although he had a wife and three young children, including an infant that was only a few weeks old, he was rarely home before 8:00 p.m. those days. Especially just then, when he was so close to cracking the problem.

He had an apparatus mounted on a workbench that looked a bit like a glass television tube placed on a table with its screen down. Some chemicals were inside it—substances he had only ever seen before on the Periodic Table. On the opposite bench, large metal probes attached to electrical wires disappeared into bulky steel canisters filled with liquid helium.

The two sets of equipment held the key to his great experiment. Inside the glass tube he was trying to create computer chips. His design relied on superconductors—chemical elements that only conduct electricity at ultralow temperatures. Helium only liquefies at temperatures of 4 Kelvins, or -269 degrees centigrade, ranking it as one of the coldest substances on earth that can be procured relatively easily. The steel vats of helium on the workbenches were being used to create a cryogenic environment.

Plumes of evaporation clouds would fill the room as the experiments were changed over. Buck knew that there were others out there trying to do the same thing: to invent an integrated computer circuit small enough and cheap enough to bring to the masses. The basic task was to find a way to create a device that could switch from an "on" position to an "off" position extremely quickly—from "1" to "0" in terms of the language of binary code upon which all computer programs depend. While the earliest computers had used mechanical switches to perform this task, scientists across the world were now racing to find better, quicker, and more efficient electronic switches. For it was only once the switches got quicker that computers would be able to start fulfilling their potential, by performing ever more complex tasks.

There were many different avenues being pursued, including the semiconducting silicon chip that eventually won the battle and drives most computers today. Yet, at the time, Buck was considered to have the scientific lead with his concept of the "superconducting" microchip. He had already won international acclaim for an earlier version of this microchip, which was manufactured using just two bits of wire wound around each other and suspended in the helium canister. Even that crude version of the device promised to become the fastest computer ever—potentially hundreds of times faster than anything commercially available at the time. But only if certain issues could be resolved. Now he was working on a more technically-advanced version.

A steady stream of newspaper reporters had trickled through his spartan little office under the dome in the main building at the Massachusetts Institute of Technology (MIT). A scriptwriter had come to interview him about turning the story of his invention into a prime-time drama. Buck and his wife had been invited to Paris for a conference that summer. MIT officialdom was also excited. A cryotron, with accompanying notes on use, was buried in a time capsule on campus in 1957 at the behest of James Killian, the president of the university, and Dr. Harold Edgerton, the world-renowned inventor of strobe lighting who built underwater cameras for Jacques Cousteau.

Yet it was in Washington, DC, that the greatest level of interest had been generated for Dudley Buck's invention. A few weeks after the Russian visit, he was due to attend a top-secret meeting of a new advisory committee for President Dwight D. Eisenhower. It was on the creation of the first generation of supercomputers for the American defense and intelligence community.

Hundreds of scientists across America were already working on the scheme, code-named Project Lightning. One of the key goals was to make Buck's new chip function. Computer scientists at NASA thought it could be useful in space. Lockheed Missile Systems and Boeing both thought it could be used as a guidance system for the newest nuclear missiles being designed.

The second incarnation of Buck's cryotron, that he was then trying to perfect, was a much more advanced device. Rather than winding wires around each other by hand, he was laying thin lines of the metals alongside each other using an electron gun. A team of more than a hundred physicists at IBM was working on Project Lightning, under contract to the National Security Agency (NSA), the newest and most obscure of America's intelligence agencies. Buck wanted to solve the remaining problems himself, however, ideally in time for his big meeting in Washington, and so he was putting in long hours. Not everyone around him grasped why he was devoted to the work.

To his students Buck was a gifted, prank-playing young professor. He was an incredible teacher who had helped out a number of less-affluent students on campus by giving them jobs in his lab to help them fund their studies. They knew he had been part of the MIT team that had designed the first computer random-access memory (RAM), an invention that helped turn computers from a curiosity into a useful tool. The full extent of his groundbreaking work was unknown to them, however.

As well as an MIT scientist, Buck was a government agent. For the previous nine years he had been working part-time for the NSA, playing roles large and small in classified defense projects such as the Corona spy satellite program, assorted missile programs, and countless schemes to build bigger and better computers for various branches of the military. He had worked as a codebreaker in Washington. Diary entries show that he was familiar with many of the Manhattan Project scientists. He had even spent time seconded to one of the most infamous intelligence arms of the CIA, which took him behind enemy lines in Eastern Europe.

Throughout his time at MIT, Buck moonlighted as one of the NSA's top troubleshooters. He was all too aware of the importance of his Cryotron chip to his superiors at NSA headquarters at Fort George G. Meade, Maryland. Since the USSR had launched its Sputnik satellite eighteen months earlier, building better computers had become an obsession of both the White House and the Pentagon.

The idea that the USSR's top computer experts would get to breeze through his lab left Buck with a sick feeling in the pit of his stomach. He was a laid-back character, an optimist. He had always been free with his ideas, telling anyone who would listen about his newest discoveries—even before they were properly patented. It had even gotten him into trouble in the past. Nonetheless, the idea of telling the Russians about his work seemed a step too far.

The trip had been arranged months in advance. He had made note of the date, writing "RUSSIANS 2 PM" in bold capitals in his diary. At the back of his mind, he chewed over how to deal with the situation.

There was little point in being too precious with information. A paper he had published four months earlier explained the experiments he was working on in considerable detail. If the KGB—the Soviet intelligence service—was anywhere near as good as it was thought to be, then Lebedev would surely have been given a copy before his trip. The paper had created quite a stir.

"The day is rapidly drawing near when digital computers will no longer be made by assembling thousands of individually manufactured parts," Buck had written in the introduction. "Instead an entire computer, or a large part of a computer, will be made in a single process."

The comment about making computers "in a single process" is a reference to the upgraded cryotron that he was making with an electron gun. What Buck was manufacturing was one of the first integrated circuits.

Lebedev and his group of Soviet scientists were originally invited to attend the conference where Buck unveiled his work, but bilateral negotiations to arrange the trip became bogged down in complications. Technical problems ensured that the first few exchanges by wire transmission were difficult for the Americans to translate. Cyrillic characters had been converted into English ones, resulting in messages that did not quite make sense. There were also transmission errors that added an extra layer of complexity. Yet the bigger problem was that the Soviet negotiators took such a long time to agree to a return visit. A year passed between the first invitation letter from the US National Joint Computer Committee and the trip taking place. The final itinerary agreed by both governments included a trip to Buck's lab at MIT.

Although Buck knew the trip had been sanctioned by the highest levels of government, he appeared a little reticent about the exchange of information. When Lebedev and his six colleagues stepped into Buck's lab on the third floor at MIT, they discovered that the demonstration they had been promised would not happen. Buck had instructed a couple of his students to remove the helium canisters and have them refilled. He explained away the problem as a badly timed piece of routine maintenance. Buck was polite and courteous, and explained to Lebedev the general principles of his work, but the Russians left with little more than what they'd known beforehand.

But Lebedev did not make a scene. In his first two days he had already gleaned more about the American computer industry than any foreigner ever had, having been given a guided tour of the IBM factory. He and his six colleagues left Logan Airport in Boston to carry on their tour, flying to Philadelphia, then Washington, and back to New York before heading home.

A few weeks after they flew back to Moscow, Dudley Buck was dead.

2
SANTA BARBARA SOUND LABORATORIES

I T WAS A COLD CLEAR NIGHT IN SANTA BARBARA, CALIFORNIA, IN THE summer of 1944. Two teenage boys had set up camp on the floor of the Shell gas station at the corner of Carrillo and De La Vina Streets.

Young Dudley Buck and his best friend Lee Meadows were determined to catch a thief. Repeated attempts had been made to break into Dan's Radio Den, a well-stocked shop selling amplifiers, speakers, and all the other radio equipment of the day.

It was only a small shop, about thirty feet long by twenty feet wide, tucked in the corner of the gas station, but the equipment inside was state-of-the-art.

Dan Foote, who owned the shop, was a close friend of the two boys. He specialized in car radios, equipping the local highway patrol cars, among others. He was one of several local electronics experts who helped and encouraged the two young radio hams—offering discounts on parts and equipment as well as weekend work in his shop. Buck and Meadows wanted to help him out; they laid a trap for any would-be thief using their radio gear.

A small speaker was bolted to the shop door, where two previous break-in attempts had been made. They had switched the wiring around to turn the speaker into a microphone, and with a long cable that ran across the forecourt hooked up an amplifier in their hideout at the opposite side. The volume was turned up high to magnify any sounds coming from the door of the shop.

As the two boys settled down for the night, they put on their heavy, Bakelite headphones and listened in. It was their second stake-out of Dan's Radio Den. The weekend before, they had climbed onto the roof of the garage and set up a listening post there. Not only was it too cold on the roof, but there were logistical issues: even if they heard the thief, they would not be able to clamber back down in time to catch him.

The new plan was much better. The sleeping bags solved the problem of the cold. Buck had somehow procured a twelve-gauge pump-action shotgun that lay by his side as they slept on the hard concrete floor. Shortly after they drifted off, a loud crackle blasted into their headphones. The trap had been sprung.

Buck reached for the gun and darted for the door of Dan's Radio Den. A man was standing with his back to the boys, carefully cutting a four-inch hole in the door with a hand drill. It would be just big enough to get his hand inside to spring the lock—and he was almost finished.

"Drop everything and put your hands up," barked the gun-wielding teenager.

The burglar jumped. He spun around to find himself staring down the gun's long barrel. As his eyes traced up to the young face whose hand held the gun, he cracked a smile. "You won't have the guts to pull the trigger, kid," he laughed, trying to call Buck's bluff.

"Yes I will!" snapped Buck. He pumped the gun to load the cartridge shell. Somehow the cartridge jumped out of the breach and dropped to the ground.

"Okay, okay, I'll go," said the burglar, picking up his tools and gradually walking toward his car, parked by the side of the building. The battered vehicle had been there for hours. It later emerged that the burglar worked Saturday nights at the Greyhound bus depot across the street and regularly parked around the back of Dan's Radio Den. The boys had not heard a car pull up, as it had been there all along.

Buck made a fresh attempt to load the gun, all the time keeping his sights trained on the bumbling burglar. Again the cartridge slipped from the gun and dropped to the ground.

"Watch what you're doing, Dud," warned Lee Meadows as he made a dash for the garage telephone and dialed the operator. "Burglary in process," he said into the receiver, repeating the line he had rehearsed as part of the plan. "Dan's Radio Den, corner of Carrillo and De La Vina."

The burglar opened the trunk of the car and threw the tool bag in the back as Buck pumped the shotgun for a third time. For a third time, the cartridge dropped out. On the other side of De La Vina Street, two off-duty marines were walking home. Meadows spotted them as he ran back from the phone and yelled for help.

They charged across the road toward the burglar. Suddenly his face fell, realizing that the game was up. As the two marines limbered up to dish out their own version of justice, a police car screeched to a halt and arrested the foiled intruder. Buck and Meadows' plan had worked, without firing a shot—or even loading the gun successfully. The two boys were sent home by the police with a pat on the back of congratulations and a warning about handling weapons.

The next day, Dudley Buck and Lee Meadows, the two young vigilantes, made the headlines of the local paper, the *Santa Barbara News-Press*. It was not the first time that Buck had made a name for himself in the local community; nor was it the first time that he had caught the eye of the authorities.

DUDLEY ALLEN BUCK was born in San Francisco on April 25, 1927, to Edna and Allen Buck. Two years later, his sister Virginia was born, and two years after that they were joined by baby brother Frank.

The family lived in an apartment at 1260 California Street, a block or so below the summit of Nob Hill—in the shadow of Grace Cathedral, the imposing neo-Gothic landmark built with money from the California gold rush. From their elevated spot the family had panoramic views of the city and San Francisco Bay; there was a park nearby where Dudley would play with his sister and baby brother. But most of the time he just wanted to build things. Every Christmas he would ask for another Meccano erector set—allowing him to build ever more complex creations.

When he wasn't building things, Dudley would take to wandering the streets—straying much farther than his mother ever realized. When he was as young as six he would take his sister Virginia down to the building site of the Golden Gate Bridge; they would stand for hours watching the thousands of men from the Bethlehem Steel Corporation bolt girders together and raise them into place, day after day, year after year. By the time the bridge towers reached their full height of 746 feet, and the bridge opened with a parade of 200,000 people on foot or roller skates, Dudley was ten years old.

Around the same time, he got a job selling magazines door to door, which gave him not only the pocket money he needed to buy more parts for his erector sets but also an excuse to keep wandering the streets. One of his favorite spots was the cable car power station at the junction of Mason and Washington Streets, one of many that kept the famous San Francisco cable car system moving. He would watch the huge cogs revolve as they pulled the loops of thick steel cable in and out of the building and under the street.

Life was good for Dudley and his younger siblings until Edna, their young Irish mother, suffered a bizarre, tragic accident. One day, at home in the kitchen, she stumbled and fell into the stove. She hit her head with such force that it caused a giant brain hemorrhage. Edna Buck was never the same again. She needed a lot of care and wasn't able to look after her family anymore. Dudley was twelve at the time.

Allen Buck spent a few months trying to juggle holding down a full-time job with looking after his wife and raising the kids on his own. He was a college-educated man with a polite turn of phrase who had an office job with the US Postal Service. Adding three children and a seriously ill wife to his workload was too much for him to handle.

The two older children, Dudley and Virginia, were sent to live with their paternal grandmother, Delia Buck, a few hours away in Santa Barbara. The decision was sudden; just a few days after they were told of the plan, Dudley and Virginia found themselves packed on the bus with their suitcases, waving out the window to Frank, their younger brother, who was left behind.

Delia Buck was a formidable woman, with a small neat frame and a piercing stare. She was of Swedish stock—the Peterson family had made their way from Göteborg to a farm in Looking Glass, Nebraska. Delia had become a schoolteacher and traveled every day to her one-room schoolhouse on horseback.

She then married Martin H. Buck, also a schoolteacher; he was a very bright man who "read for the law." They migrated to California, eventually settling in Santa Barbara. No one knows if Martin ever formally attended a law school of any kind, but he passed all the state law exams and was certified by the District Court of Los Angeles on May 13, 1905. He opened a law practice on State Street in Santa Barbara, and the family began to flourish.

They set up home in a large California-style bungalow at 1215 De La Vina Street, which runs parallel to State Street, the main business thoroughfare of Santa Barbara. The house was built in a Spanish Moorish style that was popular at the time. It had views of the Montecito Hills from the front veranda, and there was a park across the street.

Martin Buck died young, at the age of forty-nine, leaving Delia with five children (a sixth child, Hazel, had died in infancy), and the sprawling house, to look after. She had learned to do things for herself.

Delia was soft-spoken and intelligent; whenever she offered an opinion, her words were clear and unambiguous. (Many of those opinions were about the perils of alcohol—Grandma Delia led the local temperance movement.) Everyone listened to her.

By the time Dudley and Virginia were sent to live with Grandma Delia she was already sixty-two years old, and was long accustomed to life as a widow. She had learned to paint, and churned out canvases relentlessly. Each member of the family had at least one Grandma Delia original hanging on his or her wall.

If work needed to be done around the house, it was Delia who would pick up a hammer and nails and set to the work herself.

Behind the main house there was a garden with lemon, fig, and avocado trees. Then there were two small houses: a tiny guesthouse and a playhouse for the kids. A driveway ran down the middle of the yard, with garages lining either side—two dozen garages in total, butted one against the other in two parallel rows.

The garages were Grandma Delia's business. The motorcar was increasingly common in prosperous Santa Barbara, so downtown parking space came at a premium. Grandma Delia kept the family going by renting out the garages.

No sooner had Dudley stepped off the bus from San Francisco than he laid claim to one of the garages for himself. Garage number 1—nearest to the house—happened to be vacant at the time. It became Dudley's laboratory.

The windowless steel structure had a power supply but not much else. Dudley would trawl around town picking up any potential equipment he could find and drag it back to his lab. To make it clear that garage number 1 was off-limits to any visitors, he electrified the door handle.

Dudley and Virginia soon settled into the local school and got used to life without their parents, living under the rule of Grandma Delia. Their world was about to be tipped on end once again, however.

Delia, Dudley, and Virginia were in church on Sunday, December 7, 1941, when they heard that the Japanese had bombed the naval base at Pearl Harbor, forcing America into World War II. The house at 1215 De La Vina street was about to get a lot busier.

All of Dudley's uncles signed up for active duty straightaway and were shipped overseas. Their wives—Dudley's aunts Grace, Gladys, and Ruth—came home to Santa Barbara to live with their mother for the duration of the war, where they all found jobs locally.

The onset of war came amid other struggles for the family. Burt Peterson, one of Grandma Delia's farming brothers back in Nebraska, had been driven out of business. A biblical combination of severe drought, dust storms, and a plague of grasshoppers had wiped out what had once been a very prosperous farm.

Burt had grown corn, wheat, and some oats, and had reared cattle and bred horses. The Petersons were the first farmers in their county to buy a rubber-tired tractor; they also bought a big generator to supply electricity to the barn and outbuildings.

His four children had already lost their mother. After the dust covered the fence posts of the farm, the Peterson children were told to pick their most treasured possessions and jump in the car, and Burt's family also came to live with Delia in Santa Barbara. The two older sons had signed up to join the war effort, but Burt's daughter, Doris, and his younger boy, Dean, were still of school age. With Burt and his two youngest kids added to the fold, Grandma Delia's household expanded to nine. Then Uncle Ed, another of Delia's farming brothers from Nebraska, also saw his farm struck by drought; he too joined the family at 1215 De La Vina while he tried to find work.

For young Dudley, life with Grandma Delia had transformed from an existence dominated by church and school into a bubbling chaos of cousins, aunts, and uncles, all living on top of one another.

The Petersons (Burt, Ed, Dean, and Doris) blended in seamlessly with the Bucks (Dudley, Grace, Gladys, Ruth, and Virginia). At some point Dudley's parents moved down to Santa Barbara from San Francisco to join the rest of the family. Edna was still in bad condition. Allen and Edna rented their own small apartment nearby, but Dudley and Virginia continued to live with Grandma Delia, and their younger brother Frank was sent to join them.

Miraculously, Delia managed to fit everyone in. There had been an open veranda on three sides of the house, but she had glassed in two sides to create extra bedrooms. It was a full house, but a happy one.

Young Dudley was left largely to his own devices, spending much of his time in garage number 1 working on his next experiment. The small lab was primarily dedicated to the creation of pranks. One such prank was to set up a hidden microphone in his sister Virginia's bedroom, which was wired back to his garage lair; he used it to listen in on the girly teenage conversations between Virginia and her best friend Amy. Dudley had a thing for Amy. His aunts were also targeted. Early one Saturday morning he rigged a speaker just below the window of the bedroom his aunts shared. Using his amplifier, he simulated the sound of a hissing snake. The family's old dog Paddy was known to have a particular hatred of snakes. As soon as the dog heard the noise, he bounded into the bedroom, barked ferociously, and caused general mayhem. Given that Saturday morning was the only day in the week that the hardworking aunts were able to sleep in, this didn't make Dudley too popular.

The local church was also a victim of Dudley's practical jokes.

Grandma Delia was a very strict fundamentalist Baptist. As soon as the Buck children arrived in Santa Barbara, they were signed up for Sunday school, where each of them had perfect attendance records year after year. They were also sent to Bible camp every summer.

One year all of the children who had attended Bible camp were asked to share something with the congregation that they had learned. Many of the little girls recited an important Bible verse. Most of the boys held up wooden crosses they had whittled. Others had created an object or figure for the board used to tell Bible stories. Dudley, however, hauled some of his lab equipment into the sanctuary and proceeded to make a stink bomb.

As coughing fits erupted around the room and handkerchiefs were pulled from pockets to cover noses and mouths, Grandma Delia sat motionless, stunned by what had just happened.

There were dozens of other such incidents, all motivated by mischief rather than malice.

Young Dudley was a hard worker. By the time he was fourteen, he had a paper route and a job in the herb garden at the Santa Barbara Botanic Garden; he awarded himself the rather grand title of "assistant curator of the herbarium," even though his job was just to pull the weeds.

Dudley and his older cousin Dean also made money by collecting wrappers from bread and hamburger buns. One of the local bakeries offered one cent for every wrapper, so they would spend hours gathering them, earning enough to pay the forty-cent ticket price for the matinee showings at the local Granada Theater.

Dudley was an exemplary student at La Cumbre Junior High School, winning prizes and awards for everything. Virginia called Dudley Wonder Boy, and he was a hard act to follow; Virginia and Frank were known by the teachers and the principal as "Dudley Buck's sister" and "Dudley Buck's brother."

It was tough for Frank, who had a harder time with school and often fell on the wrong side of Grandma Delia's iron rule. Frank referred to Dudley as Little Jesus, even though Dudley always came to his defense.

On one occasion Dudley doctored the year's final report card that was sent home with Frank to Grandma—turning all the fails into passes by amending each F into a P with a flick of the pen. Under the section reserved for "remarks by teachers," Dudley penned, "Frank's work is a true inspiration to the whole class and his work compares with a college senior. I believe, however, that he is far too humdrummed at home. He should be given his own way more often as this encourages individualism. Sincerely, Wm. J. Kircher."

Dudley became an Eagle Scout. He also started evening classes on radio operation at the local high school, which is where he met Lee Meadows.

The two boys were perfect foils for each other. Meadows too had been forced by circumstances to move to Santa Barbara. His family came from Danville, Illinois, where his father had owned a Studebaker and Nash car dealership. The business went bust at the tail end of the Great Depression, forcing the family west to California for a new start.

Lee Meadows was the same age as Dudley. He was also knowledgeable about radios and electronics, and had designed an amplifier for a public address system while still in the eighth grade. In his spare time he was helping one of their high-school radio instructors build a similar system for use in the school.

They took evening classes together in radio electronics, which was how they came to know Dan Foote and other local enthusiasts. Dudley and Lee realized they made a good team. Dudley was the brighter of the two, and more creative, but Lee was better with his hands. They decided to set up a business together, using their electrical skills to make money.

And so Santa Barbara Sound Laboratories Unlimited was created—one of the first mobile disc jockey businesses in California, Meadows claims. "We called ourselves Santa Barbara Sound, but we weren't licensed or anything," he recalls.

The two boys bought what little equipment they could afford and scrounged the rest. From Val Shannon at Channel Radio Supply, the instructor for their evening classes, the boys acquired a Bell fifteen-watt amplifier on a rent-to-buy basis. They cobbled together the cash to buy two twelve-inch speakers, and built two more from parts borrowed from other local radio shop owners that they had met in their class. They had two microphones—one Electrolux and one Shure. Thanks to a little ingenuity, the two boys were able to cram all this equipment into Lee's Graham-Paige coupe.

It was late 1942. World War II was in full force, but it was all happening too far away to completely disrupt the flow of life in central California.

As the only two boys in school who knew anything about radios and sound systems, they soon became popular. They would set up their sound system for school dances. As word started to get out about the two young entrepreneurs, they picked up jobs providing the public address system for ballgames, dances, and just about any other event.

Then they got regular gigs at the Montecito Country Club, amplifying the bands for their Saturday night dances. According to the detailed books the two boys kept, they would get paid $13.25 for their evening's work—about $190 in today's terms.

Thanks to those high-profile jobs, Dudley and Lee were then hired by local socialite Pearl Chase to provide sound systems for her regular social events and fund-raisers. Chase was one of Santa Barbara's community pillars; the campaign to protect the town's architecture and heritage carries on in her name today.

If anyone in the area needed a sound system, they would turn to Santa Barbara Sound Laboratories. The young company was even touched by stardom: when Nat King Cole came to play the Santa Barbara Bowl, it was Dudley and Lee who provided the sound, Meadows claims. The King Cole Trio, as it was known then, had already earned a degree of fame and was about to sign to Capitol Records.

A steady income was coming from the sound business. It was not all about the cash and the glamour, however. Every time the two boys set up their equipment they learned a little bit more about the vagaries of electromagnetic fields and transformers. It led to them designing their own microphone cables to reduce the magnetic humming noise produced by their equipment.

Shortly after his fifteenth birthday, in 1942, Dudley got a part-time job with Val Shannon repairing radio receivers. Six months later he moved on to a job as a radio serviceman at Feliz Radio and Appliance at 30 East Carrillo Street.

He studied for his radio licenses along the way. In June 1943 Dudley passed exams set by the Federal Communications Commission that saw him earn a first-class commercial radio license. Although he was still just sixteen, he was hired by the local radio station KTMS the day after he passed his test to work weekends. For his twenty hours a week, Dudley earned a salary of one hundred dollars a month— about thirteen hundred dollars in today's terms.

The significant sums of cash being earned by young Dudley were mostly squirreled away for the future. He did not drink or smoke; Grandma Delia had distilled the spirit of temperance in him from a young age. He went on trips to the movies with Lee, his sister Virginia, and her best friend Amy, but Dudley got most of his kicks from his experiments.

As he started working professionally with bigger and better pieces of radio equipment, the experiments in Grandma Delia's garage became more elaborate. He built a system that allowed him to listen to his records from any room in Grandma Delia's house. It was based on a small AM radio transmitter created using a single vacuum tube. He could hook it up to his record player and broadcast the songs on a chosen AM frequency, allowing it to be picked up from a normal radio receiver. To put it in a twenty-first-century context, it was a bit like streaming music over a wireless Internet connection.

His little brother Frank and the neighbors were big fans of Dudley's self-built radio station. Though it worked very well, he kept it small and low-powered—and with good reason; it was illegal.

Amateur radio operations had been mostly shut down since the outbreak of World War II. It was part of a plot to track down spies: if the amateur signals were cut out of the equation, any signal that was not produced by the military or an official commercial broadcaster would most probably be a spy trying to communicate with his or her handlers. That was the theory, anyway.

Dudley's wireless system had attracted some unexpected attention. Pilots at the nearby air base had been disrupted on their training missions by radio broadcasts of the latest swing and big band hits. Although Dudley had been careful to avoid the frequencies used by the US Air Force, his transmitter had been unwittingly broadcasting signals on other frequencies.

The sixteen-year-old knew nothing about the flaw in his device until Grandma Delia opened the door to two agents from the Federal Communications Commission.

They dismantled the contraption in Dudley's garage laboratory, took note of his name, and gave him a stern warning about the dangers of interfering with the work of the US Department of Defense.

A few months later Dudley was plucked out of high school and sent on a fast-track training scheme for America's best and brightest.

3

THE V-12 PROGRAM

VICE ADMIRAL RANDALL JACOBS HAD COME UP WITH THE SOLUTION to a colossal problem.

The war was taking its toll on America. Many of the nation's doctors and engineers had been sent to the front line, where their numbers had been depleted. A shortage of skills was starting to hinder the war effort.

Worse than that, the production line for replacement doctors and engineers was grinding to a halt. The conscription age had been dropped from twenty-one to eighteen immediately after the bombing of Pearl Harbor, marking America's entry into the war. The young men who would ordinarily have gone on to a college education were being drafted into service before they had a chance to enroll, and those who had started their education before the war had been whisked off soon afterward, before completing their studies. The only skilled graduates now emerging from America's educational establishments were those who had already been excused from active service, mainly for medical reasons.

The shortage of students in the system created another worrying problem, with potentially lasting impact. The drop in enrollment at US colleges had been so steep that many institutions faced bankruptcy. A large part of the whole academic system was on the cusp of closing down, creating a secondary headache for the American government.

The US Navy came up with a neat solution, which it called the V-12 program. As the chief of naval personnel, it fell to Vice Admiral Jacobs to reveal the project at a specially arranged conference of 131 colleges and navy top brass held at Columbia University on May 14–15, 1943.

It was a fast-track officer-training scheme that would mix under-graduate study in a few chosen disciplines with the rigors of naval training. Empty college dormitories would be turned into improvised barracks to house this new elite force, and college quadrangles would be transformed into parade grounds.

Under the program, the US Navy would filter the best of the best of America's high-school students through a nationwide testing system. These students would then be mixed with battle-hardened marines and seasoned sailors who had been singled out for promotion to officer class but needed to improve their education.

Those accepted to the course were to be offered tuition at top academic institutions paid for by the navy, along with a salary of fifty dollars a month and a trainee officer's posting upon completion. They would study medicine, dentistry, engineering, mathematics, or more specialized subjects such as electrical engineering. There were even courses in theology, designed to increase the dwindling ranks of navy chaplains.

Unlike other students, the V-12 classes would be in uniform at all times, wake up at 6:00 a.m., and be confined to their quarters at 7:00 p.m. They would also study for twelve months of the year, rather than nine, in order to push through their degree course in record time.

All of America's academic giants, including Ivy League colleges like Harvard, Princeton, and Yale, signed up to take part. High school teachers across America pushed their brightest pupils to take the test, not least because it offered the chance to delay being sent to Germany or the Pacific.

"Gentlemen, we are about to embark on an education program that will have important effects on American colleges, on the navy and most important of all, on the lives of thousands of this nation's finest young men," Vice Admiral Jacobs told the assembled crowd at the start of the conference. "We must educate and train these men well so that they may serve their country with distinction, both in war and in peace."

The V-12 program lived up to that lofty billing. More than a hundred thousand young men were pushed through the course—allowing colleges to rebuild their finances and the navy to build up its skills base.

The scheme's output was prolific. Its graduates included everyone from future senator Robert F. Kennedy to Johnny Carson, who would go on to become America's most famous TV star. The actor Jack Lemmon and the film director Sam Peckinpah both passed through V-12, along with Warren Christopher, who would become US secretary of state. Paul Newman passed the V-12 tests but had to drop out after the navy doctors found him to be color-blind.

The V-12 project was not without its controversies. There were suspicions about whether everyone who caught on to it as a draft-dodging ploy had passed the test, or if there was a possibility that some families had used influence to keep their offspring from the front line. Its proponents argued that it was a thoroughly meritocratic affair. Whatever the truth of the situation, it unquestionably provided a route to a future for many youngsters from less affluent backgrounds who were blessed with sufficient brainpower.

Dudley Buck took the V-12 test in January 1944 and passed it easily. Five months later, right after graduating from high school, he was put on a train from Santa Barbara to Seattle. He had been assigned to study electrical engineering at the University of Washington.

Buck left charge of the Santa Barbara Sound Laboratories to Lee Meadows, who would run the business for one more year before he too signed up for a navy training course, followed by a lengthy career with defense contractor Raytheon.

By the time the train got to San Francisco, Buck had already made his first new friend. Ed Barneich had qualified for a separate training program, to become a navy pilot. After a long overnight ride on the hard train seats they arrived in Seattle to discover that they were to be roommates: the navy arranged its men alphabetically so Barneich and Buck were thrown in together.

They were posted to what had been the university's women's dormitory before the war. They were lucky; most of the other teenagers were also forced to share rooms with battle-hardened sailors.

"The navy had a sense of humor," explains Lynn Huff, one of Buck's classmates in the V-12 program. "The group there was made up half of guys coming out of high school like Dudley, myself, and the other half made of guys in from the fleet. They paired us up as roommates, the green seventeen-year-olds with the old salts. I refer to that as the education of the innocent. It was an interesting time; we grew up in a hurry."

Although the navy kept a steady eye on the curriculum for the V-12 students, the whole point of the program was that it should represent a full college education—the participating institutions were under orders that they were "expected to keep academic standards high."

Buck soon got bored of classes, however. Thanks to his extracurricular dabbling with radio back home in Santa Barbara, he already knew most of what he was being taught. He spent a lot of time helping out others, such as Barneich, who were struggling with the workload. With his spare time he soon renewed his love of practical jokes.

As part of its efforts to groom future leaders, the navy required its V-12 candidates to take courses in public speaking. Often their speeches would be recorded on 78 rpm records using a machine that cut the groove in the record right as the young students spoke. The cutting process left a long thin trail of vinyl strips. Buck realized the strips were flammable, and stored up a bagful of them. Once he had gathered a few handfuls he stuffed them into a floor lamp belonging to one of the students down the corridor. Not long after study hours began, the lamp started smoking. Buck was lucky to escape unscathed—not from the fire, but from the angry classmate who had been his victim.

That was by no means a one-off incident. Buck was forced to do five hours of drill after being caught making fudge in his room on an improvised hot plate made with some electrical wire. Undeterred, he then created a strange goopy compound that would create a small explosion when someone sat in it.

"He'd place a tiny drop of it on the chair of an unsuspecting person, who would then sit on it," explains Ken Lowthian, another V-12 student. "After a few minutes of body heat it would go off like a cap gun. It didn't cause any damage, but it sounded like a cap gun. Buck never gave up his secret. He was always a source of amusement to the rest of us."

While other students were working eighteen or nineteen hours a day, Buck could concentrate on trying to wind up one Lieutenant

Durando, the officer tasked with prowling the halls to make sure everyone was studying.

Don Balmer, one of Buck's closest student friends, claims that one particular practical joke became the talk of campus:

> We were not supposed to have any sound during study hours. Dudley thought it would be fun to take his room apart and conceal wires behind the molding. He put a radio way back in the overhead of the closet and he had it rigged so if you turned the doorknob the radio turned off.
>
> The students standing watch were in on the gag. Dudley had his radio loud enough to wake the dead. Lieutenant Durando was prowling the halls one evening. He stopped the student officer of the watch and asked, "How come you haven't gone after that radio?"
>
> "What radio, sir? I hear nothing, sir." Durando grabbed him and they went down and stood in the hall outside of Dudley's room. Other students had their doors cracked, waiting to see what would happen as the radio was blaring away. Durando asked the officer of the watch, "You don't hear that sound?" "No sir, I hear no sound, sir."
>
> Durando grabbed the doorknob and turned it; the radio instantly was quiet. Dudley was sitting at the desk doing his homework. He lumbered to his feet and said, "Yes sir, Lieutenant Durando. Is there anything I can do for you sir?" It drove Durando nuts. Later on he asked Dudley, "Just tell me how you did that."
>
> "Did what, sir?" That was typical of Dudley; he was always doing something special.

The war that they were being trained for raged on, but the V-12 students were reasonably isolated from it. In between their long days of study, there was sufficient leave to allow time for hiking trips up some of the biggest mountains in Washington, including many glacier-capped peaks over ten thousand feet high. They also went on sailing trips around the inlets of Puget Sound.

In their first few months at college, in the summer of 1944, the Soviet Union had yet to declare war on Japan. Buck and some of the other young navy cadets would canoe out to the fleet of Soviet ice-breakers docked in the bay, where they would try to strike up conversations before the sailors headed back to Vladivostok.

When the V-12 students signed out of their dorms in the evening to go to the university library, it was often just to chat up girls. The life led by this elite group was a relatively charmed one—certainly when compared to that of their high-school friends, who were now mostly in active service. The brutality of what was going on farther west in the Pacific was only evidenced by the occasional return to port of a battered aircraft carrier or destroyer, sent to the Puget Sound Naval Shipyard in nearby Bremerton for repair.

Thanks to the work of a group of pioneering scientists more than fourteen hundred miles away in Los Alamos, New Mexico, Buck and most of his classmates never experienced the horrors of World War II. The Manhattan Project scientists were nearing their first test detonation of an atomic bomb. On July 16, 1945, in the middle of the New Mexico desert, the first nuclear explosion was secretly initiated. Ten days later, Britain, China, and the United States warned Japan that it faced imminent destruction unless it surrendered soon.

For Dudley Buck, official confirmation of the nuclear technology meant just one thing: he had to try to make a bomb for himself. In between planning his practical jokes, Buck had been staying up late into the night reading what little there was to read on nuclear physics; there was only one book on the topic in the university library.

He had been telling his classmates for some time that with even a small amount of uranium it would be possible to "blow up the state of Nebraska." With hydrogen he reckoned he could make "a bigger boom." Somehow he persuaded his classmate Don Balmer to break into the university lab with him to test the theory. As Balmer remembers,

> We were in chemistry class; we had made acetylene gas and we made hydrogen gas as part of our lab courses, so Dudley figured out if we could get the hydrogen gas hot enough it would blow itself apart.
>
> So we thought that we could do this over at the university chemistry lab. We broke into the chemistry lab on a weekend, took the Bunsen burner. Dudley figured out if we had a series of concentric

tubes, he could inject gas, just the gas out of the Bunsen burner, and put some acetylene gas and then shoot into that a jet of hydrogen gas—and if that worked, we might have a big bang. So we got into the lab one weekend and we got close to the end and Dudley said, "Here we go, and if this works, we might blow up the university," and we agreed we would go ahead and do it—and of course it didn't work.

The real bomb, which had been in development since 1942, worked with devastating efficiency. On the morning of August 6, 1945, a small atomic bomb was dropped on the city of Hiroshima, wiping out two-thirds of its buildings and most of its population. Three days later a bigger bomb wiped out Nagasaki. Five days after that, Japan finally surrendered.

The end of the war brought big changes to Seattle. For starters, the university was entitled under the terms of its deal with the US Navy to reclaim its dormitories for the use of regular students, whom it expected would be returning to college. The navy students were moved to a large accommodation barge tied up close to the university, along the waterfront on Boat Street. It was 261 feet long and forty-nine feet wide at its broadest point, built to house up to six officers and 680 enlisted men. Twice a day the V-12 students would have to march back and forth from the barge to class.

Buck still found time to put his brain to work on extracurricular projects. By the end of 1945 he had designed a gas-powered internal combustion engine. Although he sent documents to patent lawyers— who were so impressed that they offered to cut their fees for the register of the design—Buck never got around to filing the paperwork.

Soon after he completed his sophomore year at the university in 1946, Buck's mother passed away. His father grew even more distant. Frank, his younger brother, asked to move to Seattle to be with Buck.

Buck, now nineteen years old, sneaked Frank onto the barge, where the teenager became a permanent fixture. Frank would have breakfast with the navy men, then head off for the day to Roosevelt High School. If the officers ever knew he was there, they turned a blind eye to his existence.

Aside from watching over his little brother, Buck had a new

hobby: salvaging electronics parts. In their spare time, the navy cadets would look across the bay at the navy ships limping back to the Bremerton shipyards from the Pacific theater. Many of the vessels had been partly destroyed by kamikaze attacks, and were mostly being stripped down for parts.

As a navy cadet, Buck had full access to the Bremerton yards. He would regularly commandeer a navy truck, drive over to the yards, and clear out as many unwanted parts as he could get his hands on: radios, cables, any scrap of electric components. Don Balmer, his accomplice in the homemade H-bomb plot, would often be dragged along for the ride.

The two cadets would strike up conversations with the young sailors who had come into port. Buck's main interest was in the radar operators. As Balmer explains,

> Radar was very new in those days, and the radar men would work and Dudley would watch them work and chat with the boys who had been in some of these big battles. Dudley was very interested in radar. When it came time to leave, Dudley nudged me and said, "Put these in your pocket," and he had me put in my pocket some radio tubes which were called lighthouse tubes that were used in this radar set. I said, "Dudley, what are these?" and Dudley said, "These are radar tubes, and I am going to make myself a radar set back at the university." I said, "But Dudley, we can't steal these things; if we get caught we'll be shot, for God's sake." And Dudley said, "Oh no, they have lots of them, and I just need a few of them." So we snuck back, and Dudley ultimately did build a radar set.

The university's radio club had its own building, and so Buck had sawed a hole in the floor and stashed his stolen parts under the floorboards. Using the equipment scrounged from the ships he built a complete, fully operational S-band radar system. There was already an antenna on the roof of the building. Along with the radar tubes he procured from the crew of one of the battered ships in Bremerton, Buck also got his hands on the official navy radar guide—which detailed how to operate and repair a radar set. From that he was able to work out how to cobble one together from scratch.

Buck started spending a lot of his time at the radio club workshop; he slung a hammock across the room and would sleep there from time to time. If the night watchmen came to check up on him, Buck would just make them coffee. After he eventually finished his project, he unveiled it to the members of the radio club, who were so impressed they wanted to brag about it.

The local newspaper was invited to see the homemade radar set, calling it a "device made of spare parts." The article said that Buck, then president of the club, had engineered and built a radar set that was "strictly from Rube Goldberg"—the famous cartoon illustrator of the day who drew pictures of ludicrous gadgets and inventions. The article went on to claim that Buck's improvised device "operates as efficiently as a standard navy radar."

The five hams in the club were also planning a project to build a television system that projected the picture onto a wall, the report added.

That was not the only gadget Buck was building. On some of their sailing trips around the bay, he had fallen in love with Vashon Island, a sparsely populated hunk of land covered with trees and berry bushes. With five hundred dollars of cash he had saved up from his radio business and navy stipend, Buck bought a heavily wooded twelve-acre plot from a Seattle attorney. He told his friends that he had a plan to build an automated berry-harvesting machine that would be able to tell ripe berries apart from green ones based on their color alone.

While that gadget did not come to fruition, his time on the island inspired a different invention: a diesel-powered machine used to rip up tree stumps. Some of the local berry farmers were struggling to clear the land, which was starting to be sold off for development. Buck sold the stump puller to local farmer James Jennings for the princely sum of sixty-five dollars.

In spite of all his entrepreneurial dealings, Buck was running short of cash by the time he entered his final year. His student days dragged on a little longer than some of his peers through no fault of his own. On the way back from one of their many student hiking trips Buck's jeep slid off the road and fell down the side of the mountain. Buck broke his cheekbone and fractured a vertebra. The thirty-pound body cast he had to wear forced him to withdraw from college for a semester.

To help make ends meet, Buck and his buddy Tom Comick found the perfect job. After spotting a notice on a bulletin board, they applied to become houseboys at the Phi Mu sorority house. Not only would the two young students have a steady income, but they'd be surrounded by girls. They also got off the navy accommodation barge: the positions came with a basement apartment in the sorority.

From their subterranean lair the two college boys could see the telephone wires running up to the dormitories. Buck knew he could just connect a phone to any particular pair of wires and listen in on any conversation he wanted. However, an arrangement as simple as that would cause disturbances on the line, and would probably be detected.

Buck delved back into his toolkit of spare radio parts and built an interface that allowed him and Comick to listen in on the girls' conversations undetected—offering them an unprecedented cache of information on the young women they were trying to seduce. And it went entirely undetected by the girls of the sorority, or the university—unlike Buck's last big prank at high school (he got in a good deal of trouble after he electrified a urinal).

Yet the time for pranks and practical jokes was coming to an end. Graduation from the V-12 program loomed, and with it came a mandatory two years of navy service. Buck had to decide how he wanted to pay back his education. Comick—who was older, having joined the college program from active service—pushed Buck to consider codebreaking, where his remarkable talents with electronics, argued Comick, would be best put to use. Very soon thereafter, Buck was thrust into the front lines of the intelligence wars.

4
SEESAW

AN EIGHTEEN-WHEEL TRUCK WAS PARKED ACROSS THE END OF AN alleyway in downtown Washington, DC. Dudley Buck, and the other young officer with him, started to panic. Their jeep was blocked in with nowhere to go.

They had no idea what was in the package they were carrying across town, but they assumed it to be important. There was rumor of spies lurking around every corner; paranoia was at an all-time high.

Buck sat behind the wheel, trying to stay calm. His partner was struggling to keep hold of his nerves.

"Move! Get out of the way!" the officer screamed at the driver of the giant rig. He got nothing more than a wave of the hand by way of response.

Pulling his pistol from its holster, the officer sprang from his seat.

"You've got five minutes to move this truck," he cried, pointing the Colt .45 squarely at the driver's face.

Flustered, the red-faced trucker fired up his engine. He stalled it five times as he shuffled the rig back and forth to make space for the two young officers—who sat revving the jeep's engine.

Washington was on edge. It was July 1948, and after a number of false starts, the Cold War had begun in earnest. Two weeks earlier, Soviet soldiers had cut all road and rail connections to Berlin. It was the Soviet Union's first open act of defiance against its World War II allies, breaching the terms of the Potsdam Agreement that had carved up postwar Germany.

Berlin, the German capital, was one hundred miles inside the Soviet zone, but the city had been split equally between the new postwar nations of East Germany and West Germany. Soviet premier Joseph Stalin thought he could lay claim to the whole city by starving out the American- and British-controlled quarters—cutting off supplies of food and fuel. The Berlin blockade ultimately backfired, but it set the tone for forty years of tension between the two new global superpowers: America and the Soviet Union.

Ensign Dudley Buck—now twenty-one and six foot one with brown hair, blue eyes, and an easygoing smile—had been posted to the navy's communications headquarters in Washington, DC. Like all junior officers, Buck had to do his share of the grunt work—incinerating classified documents or shipping sensitive intelligence to government departments or friendly foreign embassies.

When he wasn't doing that, Buck was hunting Soviet submarines, rooting out spies, and building machines to speed up the task. He had been officially assigned to work under special duty code 1615, meaning he was part of the navy's cryptology team.

This elite group of military eavesdroppers was based at Communications Supplemental Activities–Washington (CSA-W)—called Seesaw by the thousands of codebreakers, mathematicians, and engineers who worked there. The giant red-brick complex on Nebraska Avenue, in the northwestern corner of Washington, close to the border with Maryland, was where the US Navy intercepted and processed Soviet intelligence.

Housed on a sixty-one-thousand-square-meter site, the Seesaw complex had been built originally in 1875 as the Mount Vernon Seminary for Women, educating the daughters of Washington's prominent families. It schooled the children of senators, congressmen, and other luminaries—including the daughters, granddaughters, and nieces of Alexander Graham Bell, the inventor of the telephone.

In the autumn of 1942, just as America entered the war, the US Navy approached the school's aging founder, Elizabeth Somers, and offered her the grand sum of $1 million for the ivy-clad complex. Within days of the girls vacating their classrooms, the building on Nebraska Avenue was transformed into one of the most important information hubs of World War II.

Located at one of the highest points in Washington, DC, a short hop from the US Naval Observatory, it was the perfect spot from which to intercept and send messages around the country, and on to the front lines in both Europe and the Pacific.

From its new base the navy helped the British break the Germans' Enigma code, which was used to encrypt cables detailing everything from troop movements to intelligence from the heart of Adolf Hitler's empire. That was just the start, however.

Seesaw began spying on the Soviet Union before World War II had even ended, when the Russians were still allies. Now that the USSR was a clear and present danger, breaking Soviet codes and processing the intelligence had become all the more important. Operation Venona, as the Soviet codebreaking program was called, unmasked some of the most famous spies and double agents of the Cold War, such as Julius and Ethel Rosenberg, who had passed secrets of the Manhattan Project to the Kremlin. It also helped expose the British spies Guy Burgess and Donald Maclean.

Buck was posted to a junior officer's residence at Aberdeen Hall, 3415 Thirty-eighth Street NW. It did not take long for him to settle down to life in Washington, even though he was a long way from home.

His sister Virginia arrived in town at roughly the same time, having qualified as a stenographer and landed a job working for Commander John L. Nielsen in a neighboring navy building on Nebraska Avenue.

Buck often had lunch with his sister. Virginia's security clearance was not high enough to allow her access into Seesaw but she would call in advance and wait for her big brother outside the gates. Buck never breathed a word about what went on inside.

Seesaw was an eccentric place. Signals experts would play around with new Morse code combinations. Telephone gurus would practice clever ways to route calls around the world so as to avoid detection. There were linguists specializing in French, German, Japanese, Russian, Spanish, and just about every other language imaginable, some of whom would often wander around chanting out loud to practice crucial new phrases.

Banks of mathematicians sat at desks devising ever more complex ciphers, sometimes in a script that looked like hieroglyphics. There were also engineers, fiddling around with radar sets. Somehow or other they were all trying to break down Russian codes.

For some of the younger recruits, Seesaw was just a staging post on the career ladder of a navy officer—a necessary stop-off on the way to higher ranks. Others, like Buck, had been picked out especially because of their technical background.

Not long after joining Seesaw, Buck designed a new type of radar to hunt Soviet submarines. In later correspondence he described the system elusively as "special purpose pulsed radar equipment." Yet there was one thing in particular that the navy wanted him to do: invent new computers.

At the time, the word "computer" did not really have meaning outside a tight group of academics, engineers, and military scientists working on classified research projects. On the rare occasions that newspapers referred to computers, they were dubbed "electronic brains," "mechanical brains," or simply "electronic calculating machines." They were abstract devices that would be made to perform complex arithmetic in public demonstrations. The notion that anyone might want such a device in the home—or shrunk down to pocket-size dimensions and connected to a telephone—was at the time totally alien.

The concept of computers did not sit entirely comfortably with some of the veteran codebreakers, who had broken Japanese ciphers with sheer brainpower. It was from these veterans that Buck learned his craft.

One of the most famous tutors at Seesaw was Aggie Meyer Driscoll, who had volunteered for service in 1918 when a crisis of manpower prompted the navy to allow women to enlist for certain positions. She gave up her post as head of mathematics at Amarillo High School in Amarillo, Texas, and soon became one of the top cryptologists in the American military.

In World War II, Driscoll had broken many of the early Japanese codes with a notepad and pencil. Now that codes were being broken with machines her skills were becoming redundant, but she retained a formidable reputation. When everybody else had given up on trying to break a particular code, it was passed to Driscoll. She also ran a training program for young recruits like Buck.

A handful of Seesaw's veterans had spent time during the war stationed at Bletchley Park, the British codebreaking center that was set up in a stately home in Buckinghamshire, to the north of London.

Howard Campaigne, Seesaw's head of mathematical research at the time, had been America's top wartime cryptologist. He had been a math teacher at the University of Minnesota before the war, but built crude mechanical calculators as a hobby. When it looked like America might join the war, he wrote to the US Navy to offer his services. The navy made him complete a correspondence course on cryptanalysis, then hired him straightaway.

Campaigne first arrived at Seesaw on January 5, 1942, and was plunged straight into cracking Japanese codes. He had spent a year at Bletchley Park, starting in August 1944, deciphering German diplomatic cables.

Solomon Kullback was Seesaw's head of research and development at the time Buck arrived. A New Yorker, raised in Brooklyn, "Kully" had signed up to be a government cryptographer in 1930 after getting bored of life as a high-school math teacher. He had been part of the American team that broke the Japanese Navy's code Purple before he was shipped to England in 1942 to work on the German codes.

Then there was Joe Eachus, Seesaw's technical director. Before the war, Eachus had taught mathematics at Purdue University Indianapolis, where he had always liked to play around with codes and ciphers. Like Campaigne, he had been plucked for wartime service after receiving top marks in a correspondence course on cryptanalysis.

While many wartime cryptographers went on to set up companies to exploit the skills they had learned, or simply returned to academia, this small cabal of veterans had remained in service and were committed to taking on the Soviet threat. Campaigne, Kullback, and Eachus became central to Buck's career.

Although he had been processed through the navy training program at the University of Washington, Buck's studies were not yet complete. The navy sent him to night classes in advanced calculus, vector analysis, and advanced mathematical statistics at George Washington University. He used his new skills daily, especially after being tasked with helping to build Seesaw's first computer.

Captain Joe Wenger, the head of Seesaw, had become a computer enthusiast after seeing a demonstration of the Electronic Numerical Integrator and Computer (ENIAC) at the University of Pennsylvania. It was one of the first computers ever built, and much more advanced than the Colossus Mark 1 that had been built at Bletchley Park.

ENIAC was commissioned by the US Army in 1943 to solve what was then an immediate problem. Large pieces of artillery being sent into battle were supposed to come equipped with a range table that would show how far a shell or missile could be expected to travel given a set elevation and amount of gunpowder. The range table was, in essence, the instruction manual.

Every new gun design needed its own range table. Yet it took twenty man-hours to calculate each trajectory, and each gun's range table needed about five hundred different trajectories. There were not enough mathematicians capable of producing the calculations. Neither was there a stream of new students coming through, as they had all been sent to war at the age of eighteen, before they had even started to gain a sufficient knowledge of mathematics.

ENIAC cost $400,000 to build—equivalent to $5.4 million today. It occupied the Moore School of Engineering's entire basement, a space fifty feet long by thirty feet wide. It was made up of forty panels, each two feet wide by two feet deep and eight feet high, arranged in a U-shape around three walls of the room. It could execute five thousand additions per second, ranking it as the quickest computer ever built at the time.

ENIAC was not handed over to the army until February 1946, by which point the war it was supposed to be fighting was already over and there was no longer such an urgent need for artillery range tables. That did not matter too much, however: ENIAC was repurposed to perform calculations related to the construction of hydrogen bombs, and remained in operation until 1955.

The machine was unveiled to an awestruck press corps on February 14, 1946. Arthur Burks, one of the scientists who had built the machine, was given the job of demonstrating its prowess. "I am now going to add five thousand numbers together," he announced before ostentatiously pressing a button on the machine.

Headlines were made in Sunday newspapers across America and Europe the next day. The ENIAC machine became a rumbling topic of conversation, and was anthropomorphized: it was described as a mathematical "whiz-kid," a "mechanical Einstein," and a "mathematical Frankenstein."

After seeing the ENIAC demonstration, Captain Joe Wenger, the head of Seesaw, had been convinced that his codebreakers needed a similar machine. He was even more convinced after seeing the revised model, the Electronic Discrete Variable Automatic Computer (EDVAC), which could add two numbers in 864 microseconds and multiply two numbers in 2,900 microseconds. Wenger could see that processing data at those type of speeds could revolutionize codebreaking.

Wenger had a team of engineers already lined up to solve such a problem. He had encouraged a group of about forty wartime staff from Seesaw to go into business together after the war, continuing their work in the private sector. Defense budgets were being cut after the war, but Wenger told them that if they could find private capital to get started, he would hand them a fair share of juicy contracts from Seesaw. It was a way to get some of his research work off Seesaw's books, where it could avoid close scrutiny from government accountants.

Engineering Research Associates (ERA) was set up in St. Paul, Minnesota. The company hired a lot of staff who had been building military codebreaking machines during the war at an NCR computer manufacturing plant in Dayton, Ohio.

Wenger followed through on his promise. Secsaw would build a new computer named Atlas, and ERA was contracted to build it. Atlas was intended to advance Seesaw's codebreaking and data processing prowess to new levels. It was to take advantage of all the new tools inside the ENIAC and EDVAC machines.

Campaigne and Eachus were put in charge of running the project for Seesaw. With the help of Buck and the other Seesaw staff they ended up designing most of the computer themselves before passing on the design to ERA.

Seesaw decided to build a prototype in its own lab. Over just four months in 1949, Seesaw assembled a machine named ABEL. It was intended to teach the new generation of computer experts how the new technology worked and to train them on how to program the real Atlas machine when it was finally delivered.

It was Buck who got the ABEL machine up and running. He bullied Campaigne into letting him fire it up before anyone else thought it was ready. It looked like a "bare frame" to Campaigne, who laughed when he saw it. Then Buck plugged it in and it worked. It was slow, but it worked.

"Dudley Buck was a real young engineer with a tremendous amount of enthusiasm," Campaigne recalled. "He just bubbled over with enthusiasm. He was smart, but real enthusiastic."

David Brock at the Computer History Museum sees Buck's involvement in ABEL as evidence that he was already recognized as something special. "This was one of Dudley Buck's first big things in computing," he explains. "They gave Dudley, who was still very young, the job of building this machine while they waited for Atlas 1. Dudley built an analog simulator of that machine, using relays. It was the same logic, but using relay switches. He built that so that they could start to develop their programs and debug their programs on this giant, superslow simulator of the Atlas. That project must have been crazy. It must have been so huge and so complicated to build that machine—just so they could start programming."

Buck's enthusiastic research propelled him into the limelight in the codebreaking community, working with the most senior ranks of engineers at ERA and Seesaw on the Atlas machine, which was eventually delivered to Seesaw in the spring of 1950 on a heavily guarded freight train.

"I have known Ensign Buck for a little over a year and have had an opportunity to observe his work closely," wrote Lawrence Steinhardt, one of the top engineers at ERA in a letter of recommendation for Buck in early 1950. "He has been engaged in the planning and pioneer development of the electronic aspects of certain apparatus which might be characterized as digital computing equipment."

Campaigne was also asked to write a letter of recommendation

THE CRYOTRON FILES | 49

for Buck: "He is intelligent, ingenious, and quick. He is also inclined to be enthusiastic and over optimistic. He has initiative and perseverance, and is resourceful and imaginative in the designing and building of electro-mechanical equipment."

Now that the "electro-mechanical equipment" of the Atlas computer had been delivered, Buck was assigned to a new task. Although his obligatory two years of navy service were almost up, he would not see out his time in the shelter of a heavily guarded complex in Washington. Ensign Dudley Buck was about to be sent on a covert mission behind enemy lines.

5

OPERATION RUSTY

OUTSIDE THE GATES OF SEESAW, DUDLEY BUCK DREW HIS YOUNGER
sister close and whispered in her ear.

"Virginia, if anyone asks, I am going on an extended camping trip," he said. "I don't know how long I will be away. I'll let you know when I get back."

It was April 1950. There was a general feeling in the American intelligence community that the Russians were up to something. Tensions were still high regarding Berlin. America had gotten around Joseph Stalin's blockade of the city by airlifting supplies into West Berlin, and it was embarrassing for Stalin.

The mistrust of the Russians prompted President Harry S. Truman to push for more formalized integration of the Western military powers, to help persuade Stalin not to spark a new land war in Europe. The North Atlantic Treaty Organization (NATO) was born, unifying the Western powers of World War II with strategically important countries that could be vital to any future war with the Soviets, such as Iceland and Norway.

The Soviet Union was, for its part, cementing its hold on Eastern Europe, sponsoring a coup in Czechoslovakia that saw a communist government come to power.

There was now a clean divide in Europe between Western democratic states that were getting back on their feet with the aid of American loans provided under the Marshall Plan and the Eastern countries who were being bullied and cajoled closer to Moscow.

A brain drain had begun, from east to west, with many talented engineers and scientists sneaking through the border in Berlin to seek a better life. Stalin, who focused on long-term economic planning as a means to gaining more power over the West, was deeply irritated by this trend, according to CIA intelligence gathered at the time that has since been declassified.

A third world war seemed like a credible possibility, even though everyone was trying desperately to avoid one.

Buck had received orders to go to Berlin, and then into Austria—with permission to stray outside the American zones. His mission, based on order papers found among his personal effects, was described simply as "temporary additional duty concerning such matters as you have been directed to attend to."

Buck was to be honorably discharged from the navy, the papers said; he would travel into Germany on a diplomatic passport. A separate set of documents gave him permission to carry a concealed Colt .32 automatic pistol.

He would be paid a per diem to cover his expenses on the ground, the order papers said, but the US Navy would not pay this sum; it would come from some other mysterious, unnamed source. Buck had specific instructions to steer clear of military bases. It was apparent that if anyone caught him in the wrong place at the wrong time, the navy would claim no accountability for him. He was, for the time being, a spy.

As soon as Buck completed his duties, the papers said, he was to return to Washington, where he would be readmitted into the navy and sent back to his desk. In the meantime, he was to take orders from an organization called the 7821 Composite Group.

The 7821 Composite Group went by a number of different names. To the US Army it was known as Operation Rusty; to others in the intelligence community it was often simply referred to as The Organization.

It was, in short, a covert CIA operation run by a man who would later be dubbed the Spy of the Century—one of the most infamous and influential figures of the Cold War. And Dudley Buck was about to join his network.

Reinhard Gehlen grew up in the central German city of Erfurt, where his father owned a bookshop. He joined the Germany army at eighteen. He became a major in 1939, leading an infantry division into Poland as part of the Nazi invasion. From there he ascended steadily until, by 1942, he was a lieutenant colonel running Fremde Heere Ost (FHO)—a thirty-five-man crack intelligence team focused on monitoring the Russians. He was Adolf Hitler's eyes and ears on the eastern front and beyond, with sources scattered across the length and breadth of the Soviet Union.

Long before the Allies declared victory, Gehlen had started making preparations to switch sides. He had shot microfilm copies of FHO's most secret files and had them stored in watertight drums in the Austrian Alps and parts of southern Germany.

In May 1945 Gehlen turned himself over to the US Army, explaining who he was and what he could offer. He cut a deal with an army intelligence officer, Captain John Boker. Gehlen offered his files and extensive network of contacts in exchange for his liberty and the release of his top agents, many of whom were in other American prisoner of war camps inside Germany.

While postwar America was keen to recruit top German brains, dealing with anyone who was known to have been, or appeared to have been, an ideologically committed Nazi was completely taboo. Gehlen spent a year in prison in Virginia before being released in July 1946 to resurrect his operation. To win his freedom, he identified a number of members of the US Communist Party working for the Office of Strategic Services, the predecessor to the CIA.

He started to hire ex-Nazi spies, initially near Frankfurt, and then Munich. Gehlen quickly built a network of 350 handpicked agents across Germany, under the 7821 Composite Group moniker. Piece by piece, he resurrected his old network of contacts inside the Soviet Union. Eventually it expanded into a group of more than four thousand agents processing huge volumes of information. A succession of American liaison officers had no idea who Gehlen really was—he had been assigned a new name, Schneider. The information he provided was invaluable, however.

Buck reported to Gehlen on April 25, 1950, his twenty-third birthday—and just five days after he had received his orders in Washington.

Upon his arrival Buck would have found a well-funded code-breaking and signals-monitoring operation. Officially, Gehlen had a budget of $125,000 for the entire organization, but he had found a way to supplement it. Gehlen explained in his memoir how he struck a deal with one of the American colonels in Germany; he had arranged for vast amounts of surplus goods to be sent to the 7821 Composite Group from the US Army quartermaster's outpost. He then sold them on the black market inside postwar Germany at vastly inflated prices.

He also played the warring factions of the US military against one another. The declassified CIA files on Gehlen show that the air force had supplied him with "a somewhat spectacular" amount of radio equipment, and codebreaking machines, which made the army furious. Gehlen had set up a whole research lab, creating a front company called Patent and Idea Applications and Negotiation that he claimed did some legitimate consulting business.

Buck's order papers do not indicate what he had been sent there to do. Given the tensions surrounding Gehlen's position, the navy possibly wanted some simple intelligence on the nature of his operation. That seems unlikely to have been enough to justify the trip, however.

Securing raw materials may have been another motive. For several months Buck had been working on a covert project to build a new sonar system for use on America's submarines. At the time, the best radar and sonar equipment worked by passing waves through tubes of mercury. Buck had been tasked with building a system that could use fused quartz instead, which some navy researchers in Johnsville, Pennsylvania, had shown could operate at nine times the speed of the existing equipment.

The fused quartz technology worked in a similar way to fiber-optic cables, such as those used in high-speed Internet technology today. Buck was dabbling with using the cables to transmit information through a system called pulse-position modulation—now a common way to send Internet data, and a technique generally credited to two NASA scientists who came along much later.

Fused quartz looked like it could be an exciting tool in the technology battle with the Soviet Union. There was just one problem: the stockpiles of the quartz were mostly in Germany. Given Buck's involvement in the project, it is logical to assume that he would have carried back a sizable amount of fused quartz.

His bigger mission was to go on a charm offensive, however.

A FEW MONTHS before Buck set foot in Berlin, the Soviet Union detonated its first atomic bomb in a test site deep in Kazakhstan.

President Harry S. Truman, who had been reelected in 1948 on a pledge of cutting military spending, was already under pressure to reverse his stance. He was handed a sixty-page treatise by his team, National Security Council Paper 68, which became a defining tome of the era—laying out, among other things, the logic of mutually assured destruction.

The authors ran through a number of scenarios, including launching an all-out land war against the USSR while it was still weak. Eventually, however, it advocated a "rapid build-up of political, economic and military strength in the free world."

Key to the whole plan was an attempt to undermine the Kremlin's influence on the rest of the Soviet Union—to "wage overt psychological warfare" against communism. The aim was to prove America's superiority in as many ways as possible and to increase defections of key scientists and engineers.

Securing the world's best brains and getting them out of the reach of the Russians was a core strategic objective for the US government.

Buck's trip to Germany and Austria came thirteen days after Truman saw the draft of NSC 68. It also coincided with the end of a "mathematical congress" in Darmstadt, Germany, where many of Europe's top computing experts had been gathered in one place. It was a perfect opportunity to gain insight on the technology that other countries were developing.

"If they grabbed any cryptographic machine, or wanted to analyze something where they had limited time to go through it—or to analyze a device, and make an assessment of its importance—then Dudley would have been the perfect person to do that," explains David Brock of the Computer History Museum.

There was one computer expert in particular that the Americans were desperate to get to: Konrad Zuse, a former Nazi gadget designer operating in the heart of Berlin. Nobody was 100 percent sure which side of the city Zuse would end up in: the American- or the Soviet-controlled zone.

Zuse was a civil engineer with a talent for design. He graduated from a technical university in Berlin in 1935, and worked briefly in advertising. He then got a job in the Henschel aircraft factory in Berlin as a design engineer. The work involved lots of calculations, which Zuse found mind-numbingly boring. He set about designing a machine to do the work for him, working after hours in his parents' apartment.

The time-saving machine he built became his Z1 computer—a mechanical calculator with thirty thousand metal parts. He finished it in 1938, and had barely started work on the second version when he was drafted into the German army. After seeing his work, the German military gave Zuse the resources to keep working on his machines. He designed a primitive missile guidance system, and a device that made aerodynamic computations to help steer the Germans' radio-controlled Hs 293 bombs.

While the rest of the world's computer pioneers were sharing ideas, Zuse was operating in almost complete intellectual isolation. He based his second machine, the Z2, on a switching system like the ones used in early telephone networks.

The machines being built in America and Britain were much quicker, but Zuse's was more reliable. His Z3 design went a stage further: it was the first computer in the world to run from a program rather than numerical instructions punched straight into the machine.

Between 1942 and 1945 Zuse constructed his machines in a small factory in Berlin. He was selling the Z3 commercially—it is considered by many to be the second computer openly available for sale anywhere in the world. Zuse's burgeoning business was laid low by an Allied bombing raid in 1945, which destroyed the Z1, Z2, and Z3 models and all the blueprints for them. The partially built Z4 had been whisked off to a safe location.

It would be 1949 before Zuse would resume work properly on the Z4. As soon as he resurrected his lab, Zuse became a prime target for American spies, determined to lure him to their side of the great divide.

The Joint Intelligence Objectives Agency had tried to bring Zuse to work in the United States under a program called Operation Paperclip in 1949. Zuse, apparently, was not interested. IBM had bought options on his patents in 1946, but Zuse was reluctant to get too involved with Americans.

In a further attempt to woo him, the US Air Force created an unclassified job for Zuse to continue working in Berlin but under the direction of the chief of the Computing Section in the Office of Air Research. The job description was for "original work in the design of mechanical and electrical computing aids for use in the solution of engineering and scientific problems, particularly in the field of nonlinear equations."

Zuse remained a source of intrigue—and suspicion. In early 1950 he moved his computer operation to a new location near the Soviet boundary of the American Zone in Berlin. The US Office of Naval Research sent an officer to meet Zuse to find out more about his operations and make a fresh attempt to persuade him to relocate to the United States.

The report on Zuse came back saying that "the source of his funding is not known" and "not above suspicion." That was in March 1950, just weeks before Buck was sent to Germany. The Z4 machine was mostly restored, Zuse had said, but would not be fully operational until May.

Buck, fresh from the success of building his own test computer in just four months, appears to have been sent to coax Zuse to the US. Rather than send a bullying military man promising cash and status, they sent the most enthusiastic engineer they knew to lure him with tales of the resources and laboratory staff that would be made available to him.

Buck was in a much better position than even his superiors to judge what progress Zuse had made with his machines. If he could not persuade the German to move to the United States, he could at least try to see some of his machines in action and gain some intelligence.

Clearly the reception was not altogether hostile. A few weeks later, at the end of May 1950, Buck was sent back to Germany and Austria.

His eight-day mission again saw him licensed to carry a concealed automatic pistol, and he traveled in civilian clothes. So far as his sister Virginia was concerned, it was another camping trip.

On that second trip, Buck appears to have had a modicum of success. Zuse agreed to travel to America that July for an interview with Remington Rand, one of the firms building early computers in the United States, at their complex in Norwalk, Connecticut. The visit was then pushed back until September.

Zuse never agreed to move to the United States, but his firm ended up working as a contracted supplier to Remington Rand for some years, mostly through its Swiss subsidiary. As Zuse suggested in various interviews during his lifetime, the deal was eventually ended after the US authorities leaned on Remington to cut off his supply of work.

Although Zuse never revealed the names of the navy agents who came to visit him, Buck's diary entries show that Zuse came to see him at MIT later that same year, with a colleague, one H. Stacken.

Two years after Buck's trip, when Virginia was posted to Germany as part of her work with the navy, Buck would give her the name of a young man who lived near Berlin who "likes to fish and talk with Americans." She had no idea he had ever been to Germany.

6

PROJECT WHIRLWIND

Days after Buck got back from Berlin, he packed his bags and moved to Boston. His trip to Germany and Austria under the aegis of the CIA had constituted the last two weeks of his two years' obligatory service to the US Navy.

There was a growing list of corporations building computers or computer components. Many of them, such as RCA, had graduated from building radio sets and broadcast equipment. Remington Rand, the typewriter manufacturer, was investing in a big way. Burroughs, a pioneer of adding machines, was starting to get involved. Presper Eckert and John Mauchly, the two principal academics who had built the ENIAC, had also set up a business, and dozens of other university-based researchers were looking to do the same thing.

The public face of the new computing industry was Thomas Watson, the president of IBM. He had been the first to fire the public's imagination about computers when he had unveiled the Harvard Mark I, a joint effort between IBM and Harvard University, to the world's press in 1944.

Watson requested that the machine be fitted with a stainless steel casing covered in bright, flashing lights that would make it look more futuristic and appealing. It was entirely unnecessary but he hired the designer Norman Bel Geddes to produce the casing, at an additional cost of about fifty thousand dollars.

Science fiction writers were beginning to latch onto the importance of computers, thanks to these high-profile stunts. Yet most of the work in the field was still emanating from the three branches of the military—the army, navy, and recently created air force.

Buck had long had his eye on a place at MIT. For any young scientist trying to carve a path in the new world of computer technology, Jay Forrester's lab at the prestigious university was considered one of the best places to be. Forrester was assembling a team to build the world's biggest and fastest computer. He boldly named it Whirlwind in anticipation of the record speeds he expected it would achieve.

Early in 1950 Buck applied for a place at the MIT graduate school to study electrical engineering. He was rejected; MIT complained that his grades had not been high enough during his time in the V-12 program at the University of Washington. A snooty letter from MIT admissions suggested he may be better suited to finding work as a lab assistant, which could perhaps allow him to take a couple of classes; a full-time course would be too demanding for him, it said.

Buck appealed. He had been in the top quartile of his class at the University of Washington, and his class had been formed from the top 10 percent of America's high-school students. He had the grades.

After a quiet word from Seesaw, and a few high-powered recommendation letters, Buck was pushed through the MIT admission process and installed on Forrester's flagship Whirlwind project.

Buck started as a research assistant in Forrester's lab on July 3, 1950. His salary, paid out of the newly secured Whirlwind budget, was $172.50 per month. The tuition for his degree was paid for by the US Navy, and he was still in the navy reserve. The new salary he was receiving was also coming mostly from the navy, indirectly, given that it came from the Whirlwind budget. Thus, Buck was still closely connected to the military.

MIT had a rich heritage of wartime discoveries, most notably in developing radar technology. It was still involved in dozens of military projects, of which Whirlwind was one of the most costly. Forrester had been influential in securing the university's position in these projects, having been one of the first American academics to spark a fire of enthusiasm for computing technology at the highest levels of government. He had helped to show that computers could be much more than just adding machines that could spare the pencil scribblings of mathematicians.

In 1948, just before Buck started at Seesaw, Forrester published a seminal paper that raised the bar substantially in terms of thinking about what these new machines could be used for.

Forrester and colleagues' *Forecast for Military Systems Using Electronic Digital Computers* provided a series of fifteen-year forecasts on how technology could evolve—much of which sounded like fantasy at the time. The report predicted that computers would soon be able to guide missiles to specific targets, and spoke of computerized planes, machines that could transmit data around the world at high speed, and computers that could intercept Soviet atomic bombs. It also suggested that computers could revolutionize industrial processes with the use of robots. It was, in other words, a remarkably prescient piece of analysis.

To work toward these transformational goals, Forrester said, the US military would need to spend a staggering $2 billion between 1948 and 1963—about $20 billion in today's money. Forrester's proposal was labeled interesting, but expensive, and was thus pushed aside for two years. By the time Buck joined the team, however, the need to develop bigger and faster computers had become more pressing.

Whirlwind had started its life during World War II, under the code name Project Kiddy Car. It was commissioned by the US Navy, and was originally intended to be a state-of-the-art flight simulator for training wartime pilots (the US Air Force only came into being after the war, with planes the sole concern of the navy until that point).

Before it was even built, interest waned in the flight simulator idea—largely because by then the war was over, and the need to produce a high number of pilots was substantially diminished. Forrester was reluctant to let go of his funding, however, so the design was repurposed: the as-yet-unbuilt Kiddy Car computer would be developed as a computerized air traffic control system instead, the first of its kind. While those may sound like very different concepts, at their core both systems were about computing rapidly changing data about the speed and trajectory of objects flying through the sky.

The project was daring. Up to that point, computers had been devices that could be set up to run large calculations and then churn out an answer. Data would be fed in—whether it be a German encryption code, or the trajectory of an artillery shell—and a room full of men in white lab coats would wait for the answer to pop out, typically on a piece of printed tape or maybe on the screen of an oscilloscope. They often had to wait for hours. The computer would quite often break down halfway through because a valve blew up.

Project Whirlwind was different. Whether it was a flight simulator or an air traffic control system, it had to perform its calculations in real time and react to new information as it became available. To say that the technology did not exist to do the job is beyond understatement. Yet that was the whole point.

Forrester's team had been trying all kinds of new ideas to build this fantastical new machine. They were struggling to find components that could process data quickly enough to do the job. Their experiments were also chewing up a huge amount of money—about $600,000 a year.

By the spring of 1950, just before Buck's arrival, the navy was beginning to lose patience. Yet thanks to the rising threat of communism, there was a new sponsor for Forrester's cause—the Air Force.

Now that the USSR had its own atomic bomb, there was an obvious problem to address: how could a Soviet bomber be stopped from encroaching on American air space and delivering a deadly payload? George Valley, an MIT physics professor who had worked in the university's famous radiation lab during the war, had been commissioned by the Air Force to pool together a team of technical experts who could build some kind of radar machine to anticipate and block the nuclear threat. Over lunch on the MIT campus, Forrester convinced Valley that the Whirlwind system he was building could be capable of reading signals from the Doppler radar stations around the United States, serving as an early warning system for a Soviet attack.

On the back of Valley's recommendation, the air force started contributing to Whirlwind in April 1950. It was willing to put up $200,000 of the $600,000 annual bill. The reduction in spending was enough to keep the navy happy, and thus the scheme became a joint effort between the air force and the navy.

The mission was to build a computer that could track enemy aircraft, distinguish them from US air force planes, and calculate their speed and direction. Once the course of an invading bomber had been identified, exact coordinates would need to be fed to a US fighter plane that could be dispatched to intercept.

Forrester had clearly promised more than he could be sure to deliver. Yet his timing was better than even he could have realized. A few months after securing support from the Air Force, America was at war again, this time in Korea. Communist-controlled North Korea had invaded South Korea.

No one in the US intelligence agencies considered Korea a likely flashpoint. There was just one American soldier stationed at the border when the North Korean tanks started rolling in his direction. The failure was colossal. America's spy community was ripped to shreds by a series of committees, inquiries, and investigations. US intelligence had been building up its networks in Europe to deal with the Soviet threat, but now America was engaged in a land war with the communist bloc in a country many American voters had never even heard of. While intelligence officers had been training students in German and Russian, there was now a need to find Chinese and Korean speakers—and quickly.

Paranoia ensued. Seesaw was cleared to take its staffing levels back up to wartime levels; it was given new responsibilities too. As a result of the intelligence failure, codebreaking and signals intelligence was centralized in an attempt to end the infighting among different agencies. It was Seesaw that won. The organization went through several rebranding exercises over the next few years, but Buck's former commanding officers remained in prominent roles all the way through.

The immediacy of a new war ensured there was a favorable wind behind any new technical tools that could give the United States the edge in the battle against communism. Given its clearly defined purpose as an anti-Soviet defense tool, Whirlwind became of elevated importance—along with countless other computer projects across the United States. All branches of the military could see that computers were part of a future centered around nuclear bombs and long-range attacks. And the handful of individuals who knew about the technology were of prime strategic and military importance.

Buck arrived at MIT on July 3, 1950, just eight days after the North Korean tanks rolled across the thirty-eighth parallel. Thus, it was in this period of additional enthusiasm for the project that he found his feet.

His first job at Whirlwind was on its input and output systems: the ways and means of inputting data to the machine and having the results read back to the user. On a modern tablet computer or smartphone, both functions can be performed on a touch screen. In 1950, however, the method was far from settled.

Some of the more advanced computers were starting to use electric typewriters to punch numbers into the machine. Most test results were generated through some form of printout. None of that was good enough for Whirlwind—not if it was going to meet its task of identifying and intercepting Soviet bombers bringing deadly payloads to American shores at speeds of hundreds of miles per hour.

For Whirlwind to function, it needed a totally new way of operating. Buck designed a display based on a standard cathode ray tube, like those used in televisions. It was accompanied by a series of modified oscilloscopes, little different from a standard radar display. Buck built in a camera that photographed the screen periodically to acquire accurate information on the trajectory, speed, and bearing of each plane.

He then devised a system to record the radar signals being received by Whirlwind and the voice of the operator sending instructions to the pilot. It allowed the whole sequence to be captured on magnetic tape, and replayed, with the commentary that accompanied the action, as if in a movie. Those sequences could be replayed time and again without exhuasting the computer's valuable brainpower, as it was all recorded on tape. By the standards of 1950, this was revolutionary. It was swiftly tested, modified, and perfected to the point at which the military felt comfortable placing it into active service.

Yet Whirlwind had to do more than just record signals. The air force officers at the controls of the machine had to be able to pick out individual planes from the soup of information being received, and then calculate each plane's speed and coordinates. On the Whirlwind computer's screen, each plane, friend or foe, was represented by a small dot of light. It was up to air force staff to decide which dots were more interesting than others. A speedy method was needed for the Whirlwind operators to pick out the right stream of radar signals and divert Whirlwind's computing power in the right direction.

Buck sketched out a plan for something he called a target acquisition joystick. Although joysticks had been used to control airplanes since the early days of flight, no one had used one to operate a computer before. Based on the labored explanation that Buck offered for his invention, it seems that it was difficult for others to comprehend the concept, which he explained to his former military commanders in July 1950:

> A manually positioned spot on the display oscilloscope was decided upon as a convenient method of telling the computer which of several targets it is to work on. When the spot has been positioned so as to coincide with desired target on the scope face, a signal will be sent to the computer.
>
> It has been decided to experiment with a joystick type of device for manually positioning the spot with a pushbutton in the top of the joystick to inform the computer that the operator is on target. . . . Motion towards the north, south, east or west or any combination of these is done by pushing the joystick in the desired direction.

It is unclear from the various accounts of the Whirlwind project whether the joystick ever came into service. Buck complained in his notes of encountering problems with the microswitches inside his gadget. Yet he soon moved on to a different invention that solved the problem; it was a computer tool that would eventually be found in many homes in America, Europe, and Japan by the mid-1980s, and on the TV game show Jeopardy: the light gun.

The Buck light gun was a small pen-like device, with a button for a trigger, hooked up to the computer. The Whirlwind operators would just point on the screen at the plane they wanted to track, then click the button. The electron gun inside the tube could then detect the spot identified and keep track of it using Whirlwind's computing power.

It was broadly the same technology that would later be used for the gray and orange zapper device sold with the Nintendo game console. Whereas the later gadget would be used by ten-year-olds to shoot digital ducks, Buck's original machine was for targeting Soviet bombers.

IN SPITE OF the enormous challenges at the outset, Whirlwind became an overwhelming success and laid the ground for a system the US military would dub SAGE (for Semi-Automatic Ground Environment). SAGE led to dozens of colossal machines installed in huge concrete cubes at twenty-four different air bases and other military installations. Each computer took up about an acre of land. It was the immediate predecessor to the North American Air Defense Command (NORAD) missile defense system still in use today.

Within a few months of Buck's arrival at MIT, Whirlwind was starting to prove itself an impressive device. It could display friendly fighter planes with a letter F, and target planes with a T. The point at which they would intercept was shown as X.

The military's interest in the project remained intense. While the machine was still being developed with a view to air defense, Whirlwind could be utilized for a great many tasks. That in itself made it unusual: at the time, computers were still mostly considered as machines built to perform specific tasks.

For five hours a day, MIT staff were allowed to run their own programs through Whirlwind—to play with their toy and find out what else it could do. It could run about one hundred programs in that time. While it was being operated by the military, progress continued apace.

The defense chiefs seem to have been keen to brag about the creation that their endless stream of checks had spawned. Whirlwind was more impressive than anything the Russians had built. In what appears to have been a textbook lesson in Cold War propaganda, the Whirlwind machine came to be demonstrated live on prime-time American television.

On December 16, 1951, the Whirlwind machine—and Forrester, its handler—were interviewed on the CBS flagship news show *See It Now*, the main vehicle at the time for Edward Murrow, one of America's most famous newsmen. He shot to fame in World War II with his rooftop broadcasts from London during the blitz, and his coverage of the liberation of the Buchenwald concentration camp in Germany. Later he would go on to lead a campaign against Senator Joseph McCarthy's communist witch hunts—an episode of his career that was turned into an Oscar-nominated 2005 film directed by George Clooney, *Good Night, and Good Luck,* named after Murrow's wartime catchphrase.

The broadcast had the tone of a propaganda reel. Even the advertisements selected to run alongside the segment appeared to have strategic significance; immediately before Forrester's brief moment of televisual fame there was an advertisement for aluminum manufacturer Alcoa, boasting about how it was now producing four times as much aluminum as it had in 1939.

"These are days of mechanical and electronic marvels," said Murrow as he teed up his interview. "The Massachusetts Institute of Technology has developed a new one for the navy. It's the Whirlwind Electronic Computer."

Those last three words were uttered slowly and carefully: the term computer was still about as familiar to CBS viewers as the more obscure entries in a medical dictionary.

"With considerable trepidation we will now attempt to interview this machine," Murrow continued. A live video feed appeared on screen, showing the oscilloscope display of Whirlwind in the lab at MIT. It was flashing the words "Hello Mr. Murrow" with its tiny lights.

"I assume like any other piece of delicate electronic equipment, there is a human element to this," continued Murrow.

"Yes, Mr. Murrow," said Forrester, who came into shot as the camera panned out. A light suit hung from his angular frame. He was perched awkwardly on a stool, surrounded by banks of dials, gauges, and wires. He kept one hand clutched to the earpiece feeding him instructions.

Forrester proceeded to give a tour of the machine, making particular effort to point out its storage tubes, which could process data in twenty-five millionths of a second. As the camera panned across the lab, the massed ranks of electrical circuits that made the machine work were displayed.

Forrester then asked if Murrow would like to ask the machine a question. Since the navy had paid for the machine, Murrow said, wouldn't it make more sense to ask them to set a task? On a second monitor behind Murrow's desk, the navy's head of research, Admiral Bolster, had been patched in on a second live video feed.

Bolster proceeded to outline a mathematical puzzle to calculate the speed, trajectory, and fuel consumption of one of the navy's Viking rockets. It was all part of the show; the calculation had been preprogrammed into Whirlwind. It showed one bar of lights on the left-hand side of the screen, representing the remaining fuel in the rocket, and another on the right showing its speed, with the flight path charted in between. Forrester and the admiral looked incredibly smug about their work.

Murrow then asked his own question: if he had been the "Indian"—that is, the Native American—who sold Manhattan island for twenty-four dollars in 1626, how much would he have now, assuming he invested the twenty-four dollars and had received a constant rate of return of 6 percent? Demonstrating the machine's versatility, Forrester produced a preprogrammed paper tape with punched holes containing the problem. He fed it into a reading device on a giant control panel, then watched while an electric typewriter on the other side of the room typed out the answer. Murrow was clearly impressed. Anticipating the advent of accounting software, the reporting legend then asked if the machine could also work out his tax bill.

Forrester appears to have been thrown by this minor improvisation, which deviated from the script. For his parting shot, however, the lab boss produced "another type of mathematical problem that the boys here have worked out in their spare time." With that, Whirlwind started to play a tinny, electronic version of "Jingle Bells"—bringing a smile to Murrow's face.

All of these computations were straightforward for Whirlwind. The broadcast made no mention of the computer's real work for the air force, or the complex early warning system it was running. Yet the demonstration could be assumed to have served its purpose, in terms of proving technical superiority to the Soviets. The television demonstration had proven that Whirlwind was a multipurpose machine.

Its main achievement had not been visible from any of the party tricks performed for CBS viewers. It was the memory system that was driving the machine. Within a few short years, the titans of corporate America were not just copying the Whirlwind's revolutionary technology but claiming that they had invented it first. It was thanks to his exposure on this project that Buck became obsessed with trying to build the fastest, smallest computer possible. He would soon start work on his own solution to the memory problem that would see his name suddenly soar in prominence within the broader scientific community.

7

MEMORY

OVING TO BOSTON HAD POSED SOME PROBLEMS FOR DUDLEY Buck. He secured an apartment at 277 Beacon Street, just a short walk across the Charles River to the MIT campus. Yet his pay as a research assistant was barely enough to get by, due to a new strain on his household finances. Buck had become a father, of sorts.

During his two years in Washington, DC, Buck had taken charge of a scout troop attached to the Industrial Home School, a place for orphaned or homeless children and juvenile delinquents. Having had an unconventional upbringing himself under Grandma Delia's care, he could relate to their circumstances. Many of the boys had lost their father during the war.

Buck had pulled some strings to allow the boys to visit the US Naval Observatory. Scout Troop 31 even got a tour of FBI headquarters and posed for a photograph with director J. Edgar Hoover, alongside their beaming scoutmaster. He took them on hikes and camping trips, squeezing as many as fourteen boys into his 1937 Packard coupe—seven in the front, and seven in the rumble seat in the back. Often, he smuggled some science classes into their scouting curriculum too—once setting fire to a piece of parachute round the campfire while trying to demonstrate the power of hot air.

At the heart of the scout troop were three brothers, Glenn, Herb, and Bill Campbell, who lived at the school. The Campbell boys' father had died four years earlier, in 1944, after crashing a crop-dusting airplane into a barn during the night. Their mother was in and out of St. Elizabeth Hospital with mental health conditions for which there was no treatment at the time. When not at St. Elizabeth she was a waitress at the Metropolitan Hotel. She also had two girls, but they had both been adopted out.

Glenn, the youngest of the brothers, latched onto Buck. The boy had been abused by a foster parent who regularly beat him with a studded leather belt or, if that wasn't at hand, a coat hanger. Buck became a father figure to him.

Buck would occasionally let Glenn ride in his Navy truck as he went on low-level assignments around Washington. Glenn hung on every word Buck said. He even helped out Buck with a side business he had set up trading car parts. Glenn also developed a soft spot for Virginia, Buck's sister. They started to become something of a little family, spending time together on weekends.

After hearing of his scoutmaster's departure from Washington, Glenn, who was then thirteen years old, cried incessantly. Buck, for his part, fretted for the boy's future. Although he was bright, Glenn Campbell had a fondness for getting into trouble. Buck proposed a solution: he would attempt to foster Glenn, and take him to Boston.

Obviously, this was problematic. It was hardly conventional for a single, twenty-three-year-old postgraduate student to foster a teenage boy. When it was first proposed, the Child Welfare Division in Washington would not hear of the idea. Foster children needed a two-parent home, the authorities said in a sternly worded letter.

Buck had a solution, however. His sister Virginia, who was about to join the US Foreign Service, also worried for Glenn. She agreed to become Glenn's foster mother, at least initially. The Foreign Service delayed Virginia's entry into her post for a few months to accommodate the plan.

To win over the child welfare authorities, Buck also sought some high-powered recommendation letters. Joe Eachus, the wartime code-breaking hero who had worked with him at Seesaw, wrote to the Child Welfare Division, providing a glowing character reference. A second reference came from Joe Keller, an attorney with the Washington law firm Dow, Lohnes & Albertson, who had gotten to know Buck and who had been his landlord for a short while.

Soon the fostering process was in motion. The paperwork could not be arranged before Buck was due to start work, so he went on to

Boston alone. Two weeks later, a brief Western Union telegram was delivered to him at MIT: "Plan has been accepted for Glenn to go to Boston Love = Virginia."

Buck drove back to Washington the following weekend, piled Virginia, Glenn, and all their worldly belongings into the back of the Packard coupe, and took them north.

Buck initially referred to Glenn as his foster son. After just a few days Glenn asked him to ditch the "foster" part and to just call him his son. Glenn was starting to show interest in music so Buck bought him a bright red piano with white trim, and hired a moving firm to take it in through an upstairs window of the apartment.

Virginia, meanwhile, found a job as secretary to Edwin Land, the inventor of the Polaroid camera; the Polaroid headquarters were close to the MIT campus in Cambridge. The unconventional little family seemed to work. Yet paying the rent for an apartment in central Boston proved tricky. Buck also worried that the Back Bay district— now an extremely affluent, gentrified part of the city—was a place where Glenn could fall in with the wrong crowd.

He decided to use his G.I. Bill—cash benefits received by all veterans at the time—to buy a house in the suburbs. Methodical as ever, Buck took out a map of the Boston area and a protractor. He drew circles, marking various distances from the MIT campus. He and Virginia then dedicated their spare time to house hunting. It soon became clear they would have to go farther out than they had planned.

In November 1950, the family moved into 9 Birchwood Road, Wilmington, Massachusetts—a newly finished two-bedroom, one-bathroom ranch-style house being sold by a developer who needed a quick sale. Buck negotiated a price of $7,900 after securing a small mortgage from the Medford Co-operative Bank. There wasn't much left with which to furnish the place. The garden was a pile of mud with a couple of scruffy pine trees. The plaster on the walls was still drying, creating a convenient excuse for Dudley to avoid spending money on paint.

He bought two iron cots at the Army Navy surplus store in Cambridge—one for him, one for Glenn. A box-spring bed and mattress were bought for Virginia's room. They rigged some curtains from old bedsheets for the bedrooms and bathroom, leaving the other rooms bare.

Virginia, who had put her own career on hold, was unimpressed. "Sis, just think of all the different routes we can take to Cambridge," Buck told her in a vain attempt to enthuse her about the move. "Won't that be interesting—and valuable quiet time."

There was no cash left to hire someone to move Glenn's piano back out of the window of the Beacon Street apartment. So Buck got out his tools and disassembled it, piece by piece.

"Dudley was very frugal," recalls Glenn. "So Dudley took that piano completely apart and moved it to Wilmington in that '37 Packard. Somehow he got it in that rumble seat. It was amazing that he got it back together. There are a lot of strings on a piano. Most of those middle notes had three strings per hammer. It never did get back to tune. It would get close, but never exactly back into tune."

Life in Wilmington soon settled down. The Child Welfare Division in Washington was so impressed by the Buck siblings' foster parenting that, by the following spring, they asked whether there was a possibility they could also take in Glenn's sister Gretchen. Buck said he would need to be provided with a monthly allowance to support another child, and it went no further. Correspondence with Glenn's caseworker details an exceedingly suburban life centered upon school, scouts and impeccable church attendance. Glenn soon had his own bike, and a dog.

In spite of this sudden immersion into an adult world of parenting, white picket fences, and PTA meetings—and the pressures that must have arisen from coercing his sister to succumb to the same lifestyle—Buck seems to have arrived on campus rigidly focused on his work.

Although his job was related to the input and output system for the Whirlwind device, he was soon assigned to working simultaneously on the core technology driving the machine—its memory systems. As if that was not enough to keep himself busy, Buck also remained loyal to his former commanding officers in the Navy—keeping them appraised of new technologies he was discovering in the MIT research labs, and passing along research papers as they were published.

A few days after Buck strolled into the lab at MIT for the first time, he asked for permission to relay information back to Washington. Jay Forrester had written a paper outlining some of the Whirlwind team's work in the field of computer memory, *Digital Information*

Storage in Three Dimensions Using Magnetic Cores.

Having had recent access to the leading military thought in this field, Buck could see that Forrester's paper was breaking new ground. He asked if it would be possible for a copy to be sent to Joe Eachus at Seesaw. Although he was one of the most junior members of staff, Buck's suggestion led to a copy of the paper being sent to Washington.

There was nothing unusual in this flow of information. In 1950, ideas moved fluidly across academia, the military, and a handful of emerging computer giants—such as IBM and Raytheon—who earned their keep mostly on the back of government contracts. An industrial cooperation agreement bound them all together with the government and ensured that the best technology was shared. The line that marked where the state ended and the private sector began was blurry at best. Buck was operating right in the middle of the gray area.

Buck kept close to his old connections. When he took a vacation, it was quite often to go to Washington. When Eachus came to Boston, he would take his old office junior out for lunch. Buck's diary entries show that such appointments were relatively regular, even from his first arrival at MIT.

Eachus and his other wartime codebreaking friends were now running vast areas of the intelligence network, which was increasingly obsessed with building new computers. These war heroes were all mathematicians, however, who had been recruited initially for their ability to compute vast calculations by themselves. Buck brought a practical knowledge of electronics, physics, and chemistry. He could do things that they could not, and saw problems from a different perspective.

Eachus, in particular, had seen how Buck's naïve optimism ensured that he never assumed that the existing way to do something was the right way. Throughout Buck's career, his lab books were full of doodles mapping out often outlandish ideas that never got off the ground. Once, during his first year at MIT, he sketched out a theoretical design for a system that could manipulate ripples in Earth's gravitational pull as a way to communicate. Gravitational waves—which had been theorized by Albert Einstein—were only proven to exist in 2016, earning the team of a thousand or so physicists the Nobel Prize in 2017. Buck never attempted to make the sketch come to life, but it gives an indication of how his mind worked.

His arrival at MIT coincided with a period in history where no idea was deemed too wild. As the quest for the ultimate computer gained pace within the intelligence community, it soon became clear that the central problem was to build better computer memories. Projects were set up across America, all working on slightly different ways to form a zero and a one that could translate into binary code, and thus be used to program a computer. Nobody had settled upon the best way to create these switches that were needed to form the fundamental building blocks of every operation that a computer could theoretically perform.

All computers work on the same basic principle. Any calculation, command, or action—no matter how complicated—can be stripped back to a simple iteration of steps that progresses from one to the next. That brutal logic—Boolean logic, as it is called—underpins the very existence of computers.

Once ideas and concepts have been broken down into their constituent parts, they can then be translated into binary code, a numeric system that allows the simplified instructions to be expressed as zeros and ones, positives and negatives, or yes or no answers.

Whether a computer is building a nuclear missile or sending an email, at its heart it is expressing everything in long chains of zeros and ones. Although computers are programmed using different languages and codes, at the base level everything is communicated in this same way. For the computer to work, those zeros and ones need to be created, stored, and read somewhere inside the machine. That's what the computer's memory is for.

In computing parlance, these zeros and ones are called bits—an abbreviation of binary digit. Eight bits make a byte. Today's computer memories run to gigabytes and terabytes: a terabyte is one trillion bytes. Petabytes (a thousand terabytes) are even becoming part of common parlance as the "big data" revolution advances.

In the early 1950s, however, the best brains in computing science were still fretting over the quickest, easiest, and most reliable way to create and store a simple bit. The wartime machines at the British code-breaking center Bletchley Park had used electric valves, similar in principle to light bulbs, to produce the requisite ones and zeros. Their Colossus Mark 1, for example, had fifteen hundred of these valves, which could either be on or off. The ENIAC at the University of Pennsylvania, built shortly afterward, had 17,500 similar valves inside.

It was an imperfect system. Just like light bulbs, the valves got very hot and burned out relatively quickly. If just one valve blew, the computer went down.

The machines that Konrad Zuse made in his parents' apartment in Berlin used relays—electromagnetic switches that had been popularized by the advance of the telephone—to perform the same job. The switch could only be in one of two positions: zero or one. Relays were smaller, and comparatively sturdy, but much slower—even relative to valves.

The transistor, created at Bell Telephone Laboratories in 1947, was lighter, cheaper, and used less power. Yet it wouldn't be until 1954 that the device was produced on a commercial scale, and not until 1960 would an iteration emerge that could be lashed together into an integrated circuit, creating the basic microchip we know today.

Given the lack of consensus on the best way to create binary digits, and the large research budgets available, there was a willingness to listen to alternative suggestions for how best to form a zero or one.

Some researchers believed sound waves could be used to make a computer memory. The $500,000 Electronic Discrete Variable Automatic Computer (EDVAC), the second machine built by the University of Pennsylvania, and completed in 1949, trapped sound inside pools of liquid mercury as a way of storing data. At one end of the tiny pool of mercury there was a speaker, at the other a microphone. Pulses of electricity were converted to sound by the speaker which caused a vibration that traveled through the mercury. Above a certain frequency it represented a one, and below a zero. The sound would then be read at the other end of the mercury pool by the microphone. The process was repeated on a constant loop.

All the zeros and ones were stored in a perpetually cycling chain of sound waves. The problem was that if a particular piece of information was required, the machine had to wait until the next time the right segment of the wave passed through the microphone. Nonetheless, the EDVAC could add numbers in 864 microseconds.

At the opposite extreme of complexity there were punch cards. A one or zero could be determined by whether or not there was a hole in a piece of card. It was old technology—French textile mills had been using perforated paper tape to send instructions to their mechanical looms since about 1725. Charles Babbage, the inventor of the first mechanical computer, had also played around with them.

Punch cards were extremely reliable, and relatively cheap, yet they were also chronically slow. They would be used until the 1980s as a means of storing documents and processing large volumes of data. Yet they were far too slow to use as the memory for a computer's brain, certainly not for a machine like Whirlwind, which had to respond instantaneously to the threat of nuclear-armed Soviet bombers.

Then there was magnetic tape. A number of early machines in the 1940s used reel-to-reel tape recorders as a form of memory. Like punch cards, it was a technique that survived for a long time as a means of electronic archiving, but it was also too slow.

In an effort to create a faster version of magnetic tape, the magnetic drum storage system emerged. Rather than have information spooled from reel to reel, the zeros and ones were stored magnetically on the inside of a large revolving drum, not dissimilar to a washing machine. Unlike the reel-to-reel tapes, the information did not have to be read sequentially—at least not in theory.

The laws of physics were still a limiting factor. To store a sufficient amount of information the drum had to be about thirty-four inches in diameter. It had to rotate at a speed of about 115 miles per hour, or 3,600 revolutions per minute, to access data at a fairly sluggish speed of 16,600 microseconds.

At the University of Manchester, the British center of excellence for computing immediately after World War II, they developed a much quicker memory system. The Williams-Kilburn tube, named after its creators Freddie Williams and Tom Kilburn, was based on a cathode ray tube, similar to the first televisions.

Rather than creating a television picture, the electron gun inside the tube generated rows of dots representing zeros and ones. The data in the electrically charged dots were then read by a metal plate that covered the end of the tube, detecting the tiny voltage generated from the screen. The system was used in the Manchester Mark 1 machine, which went on to be commercialized as the Ferranti Mark 1, and then in a handful of other pioneering machines on both sides of the Atlantic.

It was a version of this tube that was installed in the first Whirlwind machine, the computer that had dazzled with its performance on CBS television. Yet the team at MIT was not 100 percent satisfied with the Williams tube design, so they took it apart and created their own version. Rather than use a metal plate to read the output, they decided to use a second electron gun inside the tube to read what the first one had written.

Like most things to do with Whirlwind, the tube was complicated, expensive, and groundbreaking. The tubes were all built on-site; a glassmaking factory was set up on the MIT campus. Whirlwind remained imperfect, however. The new data tubes burned out almost as quickly as the old valves did on the wartime machines. And while Whirlwind was now quick, it was not quick enough. The specification that had been agreed upon with the navy and air force required data to be retrieved in six microseconds. The tube system could only do so in about thirty microseconds—still incredibly quick for its time, but five times slower than was needed.

The problem of how to find a better, quicker memory soon became one for Buck to tackle. After his productive spell working on the display systems for Whirlwind, he was reassigned to the team trying to increase the speed and reliability of this new machine by finding a better memory system.

For years Forrester had believed it would be possible to use electromagnets as a way of storing data. It had to be possible to use electricity to create a magnetic field and turn that into a means of storage, he posited. Variations on the concept had been kicking around since the mid-1940s, but it had never been developed properly. All it needed was an electric switch of some kind to move the polarity of a magnet from north to south. Each magnet could then represent a one or a zero.

As early as November 1949 Forrester had alerted his military sponsors to the possibility that magnets could be the future. Before he had even perfected the expensive, complex storage tubes he was developing, he was telling his sponsors about how he would create their replacement.

In a letter to Captain J. B. Pearson, deputy director of the Office of Naval Research, Forrester explained,

> Recent advances in magnetic materials and the theory of solid state and electron physics make it apparent that new and improved high performance storage devices will become possible. Basic research in this field should be maintained at a modest but steady pace because one can expect that all forms of storage tubes now in use will become obsolete within a decade. Research in the first year of this program would be on certain magnetic storage and switching devices which show promise of becoming highly competitive with electrostatic tubes. Other new forms of storage could be added to the program of investigation later.

Forrester came up with the idea of running electric wire through tiny coils of magnetic wire arranged in a grid formation. Each magnetic coil would have two wires passing through it—one vertical, one horizontal. To switch the magnet from north to south would require both wires to be electrified—one wire alone would not have enough power to make the switch.

Each individual coil—or core, as Forrester preferred to call it—could be identified by the grid coordinates on the x and y axes of the mesh of wires, so each core could be easily found. He then added a third set of wires that ran diagonally across the grid, creating a third reference point.

It was this broad idea that was mapped out in the paper by Forrester that Buck asked to have sent to Eachus. Forrester had not worked out a way to make such a device, but the concept was there. Bill Papian, a graduate student in search of a thesis, was told to find the right materials and design to make this idea work. Forrester, meanwhile, went back to the giant administrative task of running the Whirlwind project.

Papian started to build a team around him. Buck was given the job of testing every type of material they could possibly find to make these miniscule magnets while also trying to improve the technology more generally.

Initially they tried to use a type of extremely fine magnetic tape called Deltamax. They would wrap it around a miniaturized ceramic bobbin, then feed the electric wire through the middle. The magnetic ribbon had to be wound by hand, so Buck invented a gadget that could be used to wind it more accurately. Eventually they were able to make a working memory cell that could hold sixteen bits of data, but it was a bit bulky.

After that they started dabbling with different types of materials. A specialist ceramics lab was built on the MIT campus just to make different magnetic compounds, all of which were tested by Buck. They baked doughnut-shaped ceramic magnets in electric ovens. Specialist ceramics manufacturers around the country were also contracted to make the tiny magnets.

A steady production line developed, producing more and more versions of the equipment. They were churning out two thousand miniature magnets a day. Each type of magnet was tested to see whether it could create a switch that flipped quicker than the last. They were strung together with different types of electrical wire in wooden grids about six inches square.

It got quicker and quicker—especially after Buck solved one of the key problems with Forrester's design. Under the original plan, each time a zero or one was read it was destroyed. As a result, the machine was constantly repeating the same work, rewriting what had just been read. Buck designed a system that allowed the numbers to be read without destroying them, meaning that the computer now had more time to process instructions.

Although the new memory was being designed for use in Whirlwind, the lab team was barred from hooking its experiment up to the prized machine until it had been sufficiently tested. The team members were forced to build a separate computer just to trial the memory system, which they called the Memory Test Computer. As they got their heads around their new inventions, the machine kept getting faster and faster.

What they had invented was the first reliable random-access memory (RAM). With RAM, data could be retrieved more quickly than it could on tape or any of the other systems that existed before. It could be accessed ad hoc; there was no need to look through the whole tape, or wait until the sound wave passed the microphone again. The acronym RAM soon stuck, and is still used today as an indication of a computer's processing speed.

On September 1, 1953, the first of these magnetic core memory units was installed in the Whirlwind machine. In a report to Eachus on the benefits of the new memory system, Buck explained that the Whirlwind machine now ran four times faster on a typical program. And it only broke down about once a week. It could perform twenty-five thousand multiplications sums per second. It could add numbers in 0.05 milliseconds—four times quicker than the wartime codebreaking machines or ENIAC at the University of Pennsylvania.

Thanks to Buck, Whirlwind was finally up to specification. Having started as a gigantic science experiment, it was now playing a part in protecting America from a Soviet nuclear threat. It occupied twenty-five hundred square feet of floor space total on two floors of the Barta Building at MIT.

Whirlwind only went out of service in 1959, when it was replaced by a fleet of machines built by IBM. All of those computers, and every other successful commercial computer at that time, used the same magnetic memory technology created by the Whirlwind team. The basics of the idea had been shared among different universities and corporations, courtesy of the industrial cooperation agreement on new computer technology that ran through Eachus at Seesaw.

Magnetic RAM would go on to become standard in all large computers of the 1960s and 1970s. It was so dependable and resilient to damage that magnetic RAM was still being used in the US space shuttle program as late as the 1980s, even though other technology had come along to supplant it. It all came down to the work done by Buck, Papian, and the rest of the team that had taken Forrester's idea and made it work.

Buck never assumed that the solution they had devised would be the end point, however. While making and testing tiny ceramic magnets for Forrester, he was continuously dabbling with other means of creating zeros and ones. He had more than just one alternative idea as to how that could be achieved.

8

FORGING BONDS

JACKIE WRAY WAS WEARING HER "BEST LITTLE OUTFIT," COMPLETE with pumps and white gloves. It was the fall of 1952; she was nineteen years old and a sophomore student at Simmons College, a women's university in Boston.

Her friend Nancy Bodenstein had bullied her into coming to a guest lecture, which was being given by Nancy's neighbor from nearby Wilmington and one of his colleagues from MIT. Jackie had been told by Nancy to get to the lecture "without fail."

The lecture began at 7:00 p.m. sharp. One of the men was young, quite handsome but boyish looking, with a slide projector under his arm. The other was older, slightly balding, and carrying a briefcase. Jackie assumed it was the older man who was there to lecture, but when Dudley Buck's name was announced it was the lean young man with the mop of brown hair who sprang to the podium as his balding accomplice loaded the slides.

Buck gave a brief introduction, then turned to the blackboard and printed the term "DIGITAL COMPUTER" in three-inch-high letters. "They are called computers, not computors," he said, as he turned to the blackboard again and carefully underlined the letter E.

Computers would eventually fly planes automatically, operate robots that could build cars, translate languages instantaneously and shoot Soviet planes or bombs out of the sky, Buck explained. All this would be achieved, he continued, with a computer that was small enough to fit in a matchbox. The whole room was gripped. It may have been a complicated subject, but Buck broke it down so that anyone could understand, not just the science majors. Jackie could not stop staring at his clear blue eyes and the perfect teeth shining through his smile.

Buck was bombarded with questions at the end, mostly from the professors. They huddled around him after he had stopped taking questions and answers from the lectern. When the fuss started to die down, Nancy made sure that Jackie got a chance to chat to the young scientist.

Jackie was smitten. Buck, too, was quickly taken in by her coy smile and bouncing blond curls. When the young computer scientist called a few days later to ask if she wanted to come see his lab, she jumped at the chance. Both maintained the pretense that they were just following up the detail of his speech with some more in-depth information.

Buck, age twenty-five, had been working for eighteen months on Project Whirlwind by this point and was well known around campus. Buck took his date on a tour of the giant building, where the enormous apparatus of the Whirlwind computer was running its calculations day and night.

Then they went to a Chinese restaurant a short walk away in Central Square. "One conversation led to another," Jackie remembers. "On our third date, Dudley casually dropped his idea that he would like to marry me. He was six years older than I and eager to settle down and eventually raise a family with lots of children. Luckily, he didn't press further. My instant internal reaction was sheer panic."

That night, Jackie told her mother what had happened, and her mother erupted in tears, proclaiming, "Oh my baby, so young."

Yet the worries soon faded. Buck's time away from the lab was so scarce that young love had time to blossom at a slightly more sensible pace. Dating was squeezed in around the edges. The young couple would meet for lunch at Durgin-Park, the famous dining room in Boston's Faneuil Hall marketplace, which did a pot roast dinner for ninety-five cents. Or they would go to the Blue Ship Tea Room, a tiny restaurant on the second floor of a building at the end of a wharf overlooking Boston Harbor. Jake Wirth's, a German restaurant in Boston's theater district, was another favorite.

When time was scarcer, Jackie would make her way to the MIT

Faculty Club. It was hardly romantic: the big round tables sat ten people. It was a place where professors from various departments argued
about all manner of things, often across multiple academic disciplines,
sketching out ideas on the paper placemats. Jackie loved to listen in.

On weekends they would take walks through the Beacon Hill district or the Boston Public Garden; at the latter they never missed a
chance to ride the famous swan boats, even though they were usually
bombarded by pigeons while doing so.

Buck had a small Plymouth sports car. They would drive to the
countryside, or sit and watch the planes taking off and landing at
Logan Airport while eating sandwiches. They went to see *Carousel*,
South Pacific, and any other musicals that came to town.

In the spring of 1953, only six months after they first met, Buck
and Jackie became engaged. At Shreve, Crump & Low, Boston's pre-
eminent jewelry store, Buck bought the ring. It was a small solitaire,
but, the salesman assured, a perfectly balanced stone. In any case, it
was all that he could afford.

Jackie had come to terms with her fiancé's odd living arrangements, including his teenage foster son. "I just thought it was wonderful that he had taken this boy in," she recalls. By this point, there
was another new lodger in the small house in Wilmington.

Virginia had finally left to take up her post with the US Foreign language school in Washington, and had then been shipped off to Bonn.
Working under John Foster Dulles, the new US secretary of state, she
was involved in arranging his many trips to the capital of the newly
created West Germany for summits with Chancellor Konrad Adenauer.

Before she even got to Germany, however, Virginia met a nice German boy. Her Pan-Am flight to Germany was grounded in Ireland for
four days by a thick fog. Georg Schick, a graduate student from
Bavaria, was also stuck. He had just spent a year in California studying at Stanford University on a Fulbright Scholarship, and was on his
way back to the University of Munich.

In those four days trapped in an airport, Virginia and Georg became close. Once their epic journey to Germany was over, they started
to meet on Virginia's leave days. Soon they were planning a new life
in America together.

In the summer of 1952, Virginia asked Dudley to sponsor Georg—or Schorry, as he was nicknamed—on his application to become an American citizen.

Schorry was twenty-five, just like Buck. Even though Schorry was said to be financially independent, Buck had to detail his own financial position in the application form. He listed his weekly earnings of $53.75 and made a declaration of assets, which showed he had $15.22 in the bank in cash, and property worth about $10,100—the (heavily mortgaged) house in Wilmington and a plot of land on Vashon Island near Seattle.

"My sister in Germany knows applicant and applicant's family; applicant has had three years of college, has studied for one year in US, desires to be school teacher," wrote Buck in the "Remarks" section of the form.

In a letter dated May 4, 1953, Virginia wrote to Dudley to tell him that "Schorry's home packing" and added, "May Day was quiet in Berlin for the Americans. We had strict orders not to be in the eastern sector for the big parade since 'hate American day' was in full swing."

Two weeks later, Buck met his future brother-in-law—and housemate—for the first time when Schorry stepped off a boat in New York City, dragging an enormous wooden trunk that was so big it barely squeezed into Buck's car. He had brought curtains, bedsheets, paintings, crockery—as well as the remains of a large fruitcake. Schorry had baked the cake with a ring inside, which he had used to propose to Virginia before he left.

Schorry enrolled in a master's program at Boston University. He also taught gymnastics to children at a Turnverein gymnasium in Lawrence, Massachusetts, a short train ride away, and worked as a singing waiter at a Hofbrau in Boston. On slow evenings in the restaurant, he marked papers for some of his professors at a rate of fifty cents an hour—until the restaurateur found out, and Schorry was fired.

On his rare evenings at home, Schorry would usually read the works of Johann Wolfgang von Goethe aloud in German by the light of the fire. The grate was usually filled with their household rubbish, or with fallen branches from the neighborhood trees. When Jackie went to visit Buck, she was quite often greeted by the unmistakable smell of kidneys cooking—Schorry's favorite meal.

Buck was away from home a lot. It often fell to Schorry to keep an eye on Glenn around the house. And then, increasingly, to Jackie too.

By the time he proposed, Dudley Buck was on a more secure professional footing. He had just completed his master's thesis on an invention called ferroelectric memory, a solid-state computer memory with no moving parts that could be developed to operate at lightning-fast speeds.

It would be more than thirty years until that technology was developed into a commercial product —what we know today as FeRAM, which drives many of the fastest tablets, laptops, and smartphones. Nonetheless, the thesis certainly brought Buck attention from the highest levels of the military and security services.

In May 1952, about six months before he met Jackie, his relationship with the military was put on a more formal basis. The army's ballistics research lab sent a delegation to visit Buck at MIT in response to his thesis. In a letter to a fellow computer scientist that summer, Buck claimed that several military labs were attempting to develop his technology. "At one laboratory, an entire two-story building (plus basement) is being devoted to ferroelectric memory research," he wrote.

Bell Telephone Laboratories, the hothouse of invention in America that had won international acclaim for developing the transistor, was very keen to hire Buck. Bell offered him a relatively handsome salary of $440 a month, but he chose to stay on at MIT, where his monthly salary as a research assistant had just gone up to $215 a month.

Around this same time, an old contact from Washington got in touch. The organization that had subsumed Seesaw and was about to be rebranded as the National Security Agency (NSA) wanted to hire him as an "expert"; he would be paid at a rate of thirty-five dollars a day—about three and a half times more, pro rata, than he was paid by MIT.

The demands on Buck's time would not be sufficient to force him to give up his studies or research, and this was clearly an attraction. Yet the chain of correspondence suggests Buck did not get much say in whether or not he took the job. He was already being passed through the application process before he got clearance from MIT to take on the extra work. He sent a somewhat bashful memo to his direct line manager asking if it was permitted for a junior research assistant like him to take on such high-profile external consulting work.

While Washington's interest in Buck seemed to raise eyebrows with his professors at this time, no one at MIT ever said no to a government agency—and especially not one that worked so closely with the university and provided so much of its funding.

In 1952 the NSA was so secret that no one would even confirm its existence. It was nicknamed No Such Agency among other branches of the military for the comedic frequency with which US officials denied knowledge of it. It first got a fleeting mention in a US government organizational chart in 1957, but the letter from President Harry S. Truman sanctioning its creation remained classified for much longer.

"In the first half of the 1950s, the NSA was the prime consumer of digital electronic computers," explains David Brock, director of the Computer History Museum in Mountain View, California. "It was foundational for the digital electronic computer industry in the United States. It was the first customer for all the digital electronic computers made by mainframe manufacturers. They were buying the first, and most expensive, units—often several of them. In that, the NSA helped to make further units cheaper and so promoting their use in science and engineering, then soon into business and bureaucracy. The NSA was extremely important."

Although the White House was still denying the NSA's creation, Buck was receiving letters from the organization on headed notepaper—and claiming back travel costs by attaching receipts to official government expense forms bearing an NSA rubber stamp. Years later, the carbon copies of those NSA travel receipts were among some of the extraordinary items found among his notebooks and diaries. The receipts place Buck in some interesting places, at some interesting times.

For example, around the time that America started testing hydrogen bombs, Buck returned from one trip to the West Coast with a collection of extravagant seashells that he explained were only available in the South Pacific. Had he witnessed the test explosions firsthand? Or had his trip simply been a coincidence? It is clear from Buck's notes and diaries that he knew the individuals involved in the tests.

Correlating his travel plans with his diaries and lab books show a similar pattern emerging over a period of years. Frequently, at times of geopolitical crisis or intrigue, Buck would travel to Washington or to a military installation around the country—and file an expense claim back to NSA headquarters at Fort George Meade, Maryland. Then shortly afterwards, diary entries show, a meeting would take place at MIT between Buck and another scientist from a different institution or branch of the military that we now know to have been involved in a particular classified project of note. Sometimes the meetings were listed in his diaries under their military code names. Sometimes just the name of the individual. In some of those instances, lab book entries show that he devised experiments to solve particular problems. In other cases, he seems to have been used simply as a sounding board. For it was already clear from his achievements that Buck's brain was wired slightly differently from even those of his respected peers.

"Dudley Buck was seeing stuff well into the future," explains Brock. "People are still grappling with things today that he saw sixty years ago. Some of it is quite far out there—neuromorphic computing, for example. He was trying to build a computer, in a three-inch cube, to replicate the human brain. People are only now getting back into that."

AFTER BEING REHIRED into a formal role in clandestine military operations, it seems that Buck's first task was to be brought up to speed with what had been occurring in his two years as a civilian. He was asked to attend a science conference convened under the auspices of the NSA at government research labs in Corona, California, in May 1952. The order to attend came before he was even officially sworn in to his new role.

It was an important affair. Among the other scientists attending was Seth Neddermeyer, one of the top physicists from Los Alamos, New Mexico, home to the Manhattan Project—where America's first nuclear bombs were conceived and created. Neddermeyer was twenty years Buck's senior and a revered figure in the science community.

Also attending was Joseph M. Pettit from Stanford University, who had received the President's Certificate of Merit for work he had done with radar during World War II. Pettit brought along his colleague Lester M. Field; the two men were already developing a highly advanced radio amplifier called a traveling wave tube that would be used in the communication system of Telstar satellites.

Then there was Zoltan Bay, from George Washington University, a Hungarian professor who left his communist homeland to come to the United States in 1948. Though he had worked with the famous mathematician John Von Neumann and other Washington-based computer experts on their early computers, Bay was most famous for successfully devising a way to bounce radar signals off the moon, a technique he had perfected in his lab in Budapest in 1946.

Buck was to spend three weeks in California mingling with these big brains. It was not only his first mission as an NSA expert but also one of his first trips back to his home state since he had set off on the train to Seattle as a wide-eyed teenage navy cadet. The trip would be a chance to catch up with his Grandma Delia, his little brother Frank, and the rest of the family in Santa Barbara, where his return would catch the attention of the local newspaper. He brought Glenn with him to meet the family. A clipping in Buck's papers showed that the *Santa Barbara News-Press* told its readers that the town's prodigal son had returned for a "summer job" at the National Bureau of Standards lab in Los Angeles, adding, "The twenty-five-year-old Santa Barbaran will return to MIT this fall to work on his doctorate."

Buck's trip was a little less innocent than this cover story suggested, however. The invitation had come principally from Joe Eachus, Buck's main mentor from his time in the navy. Eachus was now the chief of the NSA's Analytical Equipment Machines Division in the Office of Research and Development. Loosely speaking, he was the NSA's top computer scientist. Howard Campaigne, the other wartime codebreaker who had taken a shine to Buck during his time at Seesaw, held a similar post to Eachus in the still-new NSA, but with a slightly different title. The two men flew out to California to see Buck, spending a day with him at what was described as a "mathematical symposium" at the University of California–Los Angeles.

No one else attended this "symposium" other than a professor from UCLA, it appears. At this time, the NSA was at war on two fronts: laboring over the Soviet codes, and battling the ever-changing demands of the Korean War. After his briefing, Buck was sent to the secluded mountain laboratories of the National Bureau of Standards in Corona, to join the other assembled brains.

The Corona lab was different from the rest of America's computer facilities in one key respect: unlike its peers, the Bureau of Standards avoided developing its own components, preferring to perfect technologies that others had patented. Yet it was also developing high-speed computers. It had built a machine called the Standards Western Automatic Computer (SWAC) that went into service in 1950. SWAC briefly held the title of the world's fastest computer, until MIT's Whirlwind machine came along.

Buck became a regular visitor over the years that followed, handing over information about what he and his colleagues were developing as well as advances he had learned about elsewhere. On that first trip, however, the secluded lab in the Santa Ana Mountains was mostly just playing host to this odd congress of scientists. Buck, Neddermeyer, Bay, the two Stanford professors, and a handful of other scientists with links to various government programs spent a fortnight holed up together in Corona. Part of Buck's role was to send notes on the discussions to Colonel George Campbell in Washington. He distilled two weeks of discussions down to a two-page memo.

Based on the account he sent back, which was found among his papers, a large part of the meeting focused on the magnetic core memories that Buck and his colleagues had developed at MIT, which was now driving the mass roll-out of computers. Some slightly wilder ideas also emerged, however.

The Korean War—which began in June 1950 and lasted until July 1953—had raised the stakes for America's scientists. Throughout the conflict, America's upper hand rested partly on the threat that it could do what it had done in Japan a few years earlier: drop the bomb. The technology was still evolving, and rapidly.

Four months after this meeting in the Californian mountains, America used technology devised by Neddermeyer to detonate the world's first hydrogen bomb—a device up to a thousand times more powerful than the atomic bombs that had exploded in Japan. Ivy Mike, as the bomb was called, represented a significant symbolic strike against the Soviets. Film footage of the explosion was broadcast around the world after it had been approved by the censors.

The Ivy Mike detonation used a particular form of hydrogen: one thousand liters of deuterium, kept in liquid form by a giant cryogenic freezer that generated ultralow temperatures. Deuterium had only been discovered in 1931. It is also known as heavy hydrogen; it is heavier and less stable than hydrogen. Deuterium is found in trace amounts in the world's oceans, and it is generally believed that all the deuterium found in nature has been present since the Big Bang. Water containing deuterium is referred to as heavy water, or deuterium oxide.

The Nazis dabbled with heavy water during World War II in their attempts to build a nuclear reactor and as part of their attempts to produce an atomic bomb. A sizable Allied sabotage plot was undertaken to shut down the Germans' secret heavy water factory in Norway. After the war deuterium came back into fashion as a bomb-making material. It was of strategic importance, therefore, but it was also difficult to procure.

The first American experiments with deuterium took place at the National Bureau of Standards' other facility in Washington, DC, where a new state-of-the-art physics lab turned it into a liquid form of hydrogen by freezing it at ultralow temperatures.

During the two-week conclave in Corona, it was resolved that procuring more heavy water would be a wise idea. In the months following the seminar, this subject became a homework project for Buck and Lieutenant Commander M. Scott Blois, a navy scientist and former colleague from Washington, DC.

Heavy water is about 10.7 percent denser than normal seawater. Normal tides and ocean currents stop it from sinking to the bottom of the sea, however. That was the accepted wisdom, in any case. Blois and Buck posited that in the deepest parts of the ocean, at depths of ten thousand meters or more, higher quantities of deuterium may be present. As they theorized more formally, they concluded that sub-sea pressure could mean that seawater in deepwater trenches could be

11.7 percent heavier than normal seawater. Thus, it was possible that sizable quantities of bomb-grade hydrogen may be simply lurking in the mud of the world's deepest ocean trenches.

Their interest was not just with a view to building bombs. Buck, Blois, and the rest of the experts assembled in Corona hypothesized that deuterium could also be used as a basis for high-speed computer memory.

Buck outlined the concept in his notes, describing it as "nuclear induction memory." In certain states, this highly explosive nuclear fuel could be moved and manipulated by magnets, and used to store images. Buck and Blois formed the view that the same idea could be applied for the ones and zeroes; a magnetic pulse sent into a deuterium store could be used to fashion ones and zeros as a computer memory.

The physics behind the idea had been understood for a while. The concept is central to the magnetic resonance imaging (MRI) scanner used in hospitals everywhere to survey and analyze the human body.

Buck's debrief for Colonel Campbell described the deuterium idea as "quite exciting from a computer memory standpoint." He added that there was "considerable effort" going into this research in other labs across the country, and that the group of scientists that had convened in Corona would "keep in touch with the developments."

Blois soon procured samples of deuterium from the Tonga Trench and the Philippine Trench—two of the deepest spots in the world's oceans—after tracking down a local marine biologist who had been aboard a Danish research ship a year earlier.

Some brief experiments proved that their theory was right: the mud had a much higher concentration of deuterium than did normal seawater, making it much easier to turn into a nuclear bomb, a computer memory, or both at the same time: a bomb that was controlled by a computer built out of its own payload.

Blois wrote an excited letter to Buck, explaining that their theory had received a favorable response in the Office for Naval Research (ONR). "I sent the proposal to ONR," he said. "They talked it over with physicists and oceanographers there and stated that they thought the idea an interesting, a promising one and suggested that it be followed up. I therefore proposed an approach which is compatible with my main job here and will see what happens. . . . I have mentioned you in my letter to ONR as the co-author of the proposal, so you are involved to that extent."

On his return to MIT, Buck drew upon the resources of various departments at MIT to assist with the analytical work for the "nuclear induction memory." Project Galathea, as it was christened, rumbled on for a few more years before it became overtaken by more pressing laboratory work. While Buck and Blois experimented with deuterium as a computer memory, scientists elsewhere in the United States were trying to do the same thing with everything from lemon Jell-O to a particular type of hair gel called Wildroot Creme Oil. The hair oil responded to a pulse or radio signal when they were in a strong magnetic field. Such was the desperation in the search for new computer components, and the willingness to entertain any possible alternative means of creating a switch that could flip from an on position to an off position at the fastest speed possible.

Buck and the other scientists who had gathered at the summit in Corona became a forum of sorts within the NSA named the Working Group on Ultra High-Speed Computer Circuits. Ultimately they were working toward building the first generation of the NSA supercomputers.

9
PROJECT NOMAD

D
UDLEY BUCK STARTED SHUTTLING AROUND THE COUNTRY FROM one classified project to the next on behalf of the NSA. Part of his role was to straddle the various top-secret computer projects commissioned by the agency to make sure that every organization involved was aware of the latest techniques. Some of those institutions were universities or government labs like the one at Corona. Others were commercial firms like Bell Telephone Laboratories, Engineering Research Associates, and IBM.

Now that he had been brought back into the military fold, Buck's orders came directly from Solomon Kullback, another ex-Bletchley Park codebreaker, who outranked Campaigne and Eachus as the NSA's chief of research. Kullback was something akin to Q in the James Bond films and rose to the rank of colonel before he retired from spycraft in 1962.

Kullback had been one of the US Army's first dedicated codebreakers—the third employee hired into the army's Signals Intelligence Service, as it was then called. Cracking codes remained a key concern for him, as it had been since he had broken the Japanese military ciphers in the 1940s.

Although computers had shown potential for lots of other uses, it was in coding and decoding secret messages that computers had established a proven track record of performance. Through Kullback, Buck became involved in developing an encryption system that would evolve into a device called KW-26—soon the standard coding machine for the US government and later for most NATO countries.

The other long-standing perception of where computers could be of greatest use to military society was in translating languages. Once a codebreaking computer had cracked a code, the messages still had to be translated from the original language into English before its importance could be analyzed; that still required human intervention. If thousands of pages of decoded messages had to be sorted through, from listening posts all around the world, it could take ludicrous amounts of highly skilled manpower.

This was a problem that had been recognized by America, Britain, Germany, and the Soviet Union during World War II; all of the military powers attempted to build translation machines with varying degrees of success. Even twenty-first-century artificial intelligence systems struggle to master the nuances of natural language translation, yet it was viewed as a logistical challenge at the time.

Thanks to the Korean War, American intelligence services were now having to deal with more languages than they had before. This made the demands of breaking Nazi messages a few years earlier seem easy by comparison.

The North Koreans were backed by the Chinese, while the Soviet Union remained the perennial threat; this created a demand for knowledge of three languages. The sheer volume of spying going on in and around Berlin, meanwhile, ensured that there was still demand for German speakers and translators.

Postwar tensions in the Middle East were also starting to bubble up—events that would later lead to the Suez Crisis and the Six-Day War. Languages were in high demand, so the NSA was under pressure to find a way to speed up the translation process. It was not humanly possible to keep up with the flow of information being generated.

A conference was organized by Yehoshua Bar-Hillel, a prominent philosopher and linguist at MIT who was at the vanguard of a group who believed that any language could be translated by machine if given enough data. The International Conference on Machine Translation convened at the MIT campus, where the all-singing, all-dancing Whirlwind computer was already in operation. Attended by eighteen of America's top linguists, mathematicians, and computer scientists, the conference ran from June 17 to June 20, 1952.

Bar-Hillel would later team up with a young philosopher of soaring importance, Noam Chomsky. "Professor Bill Locke suggested we use computers to do automatic translation, so we hired Noam Chomsky and Yehoshua Bar-Hillel to do it," Jerome Wiesner, the former dean of MIT, recalled in an interview for the university about the demolition of the building where the two men worked. "It didn't take us long to realize that we didn't know much about language. So we went from automatic translation to fundamental studies about the nature of language."

Buck was sent to Bar-Hillel's conference as an observer for the NSA. MIT's delegation was led by Wiesner, who went on to become science adviser to President John F. Kennedy and later acquired a degree of infamy during the Watergate scandal: leaked papers revealed he had been one of twenty academics denounced as hostile to the administration of President Richard Nixon.

Buck was one of only two attendees who had yet to achieve a grand title, the other being James Perry, a research associate at MIT. Everyone else in attendance was director of an Ivy League university department or ran a government laboratory. As was customary at such events, there was also one representative from IBM and one from the RAND Corporation.

The only foreigner was Andrew Donald Booth, a British computer pioneer then working at Birkbeck College in London. Booth—originally an explosives expert who had invented the magnetic drum computer memory—would have been known to many of the individuals due to a postwar spell he spent at Princeton University working with American mathematician John Von Neumann on his early computers.

Some of the sessions were open to the public, and the assembled experts took questions from the floor. Buck wrote an extensive report for the NSA about what was discussed in terms of the agency's particular areas of interest.

In both the public and private sessions there were some attempts to tackle the eternal philosophical questions about language and the common syntax and structural elements shared among different groups of languages. They also dabbled in the definitions of machine intelligence, as mapped out by Alan Turing only a few years earlier in his work at Bletchley Park; the famous Turing Test of divining whether a machine or a human being is answering questions from the other side of a screen.

Yet most of it was a fairly pragmatic attempt to scope the problem of deconstructing languages to make words and sentences readable by machine. How would machines cope with scientific jargon? What about idioms and odd turns of phrase? Stuart Dodd from the University of Washington in Seattle suggested the solution may lie in standardizing the English language.

"In English, at least, regularizing leads only to a certain quaintness of expression somewhat similar to the sentence structure employed by the Quakers," wrote A. C. Reynolds of IBM in his account of the session. "No attempts have been made as yet to regularize languages other than English, but at least for the Romance languages it seems on first view that such regularization can be accomplished."

Bar-Hillel himself explained that the United Nations was processing about 175,000 pages of translated documents every year. To keep up with the flow of Chinese and Soviet messages being generated solely as a result of the Korean War, the NSA would need to process more than a million words per day. "I used the phrase 'if a human being can do it, a suitably programmed computer can do it too' more than a dozen times during the conference," Bar-Hillel wrote in a 1964 paper recalling the event.

Bar-Hillel spent the next twenty years researching computers. By the time of his death in 1975, he had concluded that it was more or less impossible to build a machine that could translate accurately and consistently using proper syntax and grammar. He handed over his work to Chomsky, whose fame has subsequently outstripped that of his professor.

After Bar-Hillel's conference, a whole new area of computing research was spawned, with programs set up across the United States to investigate this new form of artificial intelligence. As the drive for machine translation progressed, it fell to Buck to keep Washington up to speed on developments—and contribute his own ideas.

There was an assumption that building these translation machines would also mean finding a new way of building a computer. The invention of the electronic transistor in 1947 had led to a new boom in computing, replacing vacuum tubes as the driving force of computers. The next stage of development, it was assumed, would involve some yet-to-be-discovered device.

"The transistor was smaller, used less power," explains David Brock of the Computer History Museum. "The NSA and the aerospace people ate that up. They wanted transistorized computers. Even the people who made the transistor were thinking, 'Okay, what's next, what's the next device?' Clearly the transistor was better than the vacuum tube. Obviously, there's going to be something that's better than the transistor. So people were looking at all kinds of things."

Infused with the knowledge of the MIT linguistics department, the electrical engineers set to work in attempting to build translation machines. On November 28, 1952, Buck wrote to the US Department of Defense claiming that a team at MIT was making advances in its efforts to translate Russian by computer: "Work is progressing here at MIT on the short-range word-for-word translation program under the direction of Professor J. W. Perry. The chosen language is Russian. Enclosed are sample translations and a discussion of the machine simulation processes. The work is primarily aimed at automatic indexing and abstracting."

Washington took the view that this was a task that merited a large-scale, dedicated contract. For any translation machine, the problem was how to index and file a whole language so that it could be easily retrieved in microseconds. It was a problem of computer storage, therefore, as much as a cryptographer's or linguist's intellectual game.

The NSA had commissioned Raytheon, the defense and electronics giant, to work on this problem at its labs in Waltham, Massachusetts. Project Nomad, as the program was named, was an attempt to create a filing system that could be used to convert a human language into a set of searchable data points. Roughly once a month, Buck was sent to Raytheon to check up on progress. Yet his dispatches back to Washington told a story of a constantly changing plan with little material progress.

As Campaigne, one of the recipients of those dispatches, explained in a 1983 oral history interview conducted for the NSA,

> Nomad was a flop. We never got anything much out of it. . . .
> We let a contract to Raytheon, up in Boston, to do this and we
> had a bad experience with them. Their proposal sounded very
> imaginative. It really sounded good. The minute we signed the
> contract everything changed. They no longer were imaginative.
> They were picayune about changes.
>
> Anything that we mentioned was a change in contract,
> change in scope, change in price. I suppose that's what they
> did—they took the imaginative man off and put him on some-
> thing else and they put some other person on it. He was a fiscal
> type person.

The venom in the phrase "fiscal type person" says a lot about the
attitudes of the day. Budgets were there to be blown; the mission was
to get ahead of the Russians.

Project Nomad ran for two years before it was ditched in June
1954. The NSA concluded that it was so far off track that the system
was likely to be obsolete by the time it was finished: "When the books
were closed, the contract was shown to have cost the government approximately $3,250,000," the agency wrote years later in a review of
its computer programs.

Nomad's colossal budget was not entirely wasted, however.
Buck's notes and diaries show that he managed to get the Raytheon
team to do a lot of experimentation with some other computer systems he was developing. They also dabbled with the ferroelectric
memory chip that had formed the basis for his thesis.

On his trips to Raytheon, Buck found himself addressing teams
of 450 or more engineers on the finer points of his newest inventions. The engineers were then sent off to do his research as part of
the taxpayer-sponsored project.

As for the problem Nomad was supposed to solve, Buck later
came up with a solution on his own. He created a system called
content-addressable memory, a filing method that allows a big data-
base of information, such as the entire contents of a language, to be

searched in record time. It was developed as part of the work on the Whirlwind machine.

Content-addressable memory is one of the systems that allows big data computer systems to sift through vast amounts of information and pull out only the relevant parts. Rather than work its way through every piece of data sequentially, it can search them all at once.

"A normal addressed memory is like a list of telephone numbers," explains David Brock of the Computer History Museum. "And you call those numbers to see who is home. If you are searching for someone, you would go through all those memory locations and call all those numbers until you find the right person. Content-addressable memory is a way to call all the phones at one time and ask, 'Is Dave home'? It registers when someone says, 'Yes, I am here.' It's really fast."

Of all Buck's inventions, content-addressable memory was the one that—thus far—has had the most direct impact on the age of the Internet and modern computing. At the time, it was just another solution to just another problem dictated by the military.

"Content-addressable memory continues to be a thing," notes Brock. "It is one arrow in the quiver of modern data scientists. There's a lot of artificial intelligence that does not use content-addressable memory. But it's still in use. The first patents on that are the work of Al Slade and Dudley Buck. They built one for the NSA. I haven't seen earlier stuff."

Inventions were coming thick and fast across the industry. With so much money being spent, and so many brains dedicated to the cause, the American computer scientists were innovating at a ridiculous pace. What was happening under the guidance of the NSA was substantially ahead of anything available commercially. In late 1951, Computer Research Company in Hawthorne, California—one of the early computer component manufacturers—was advertising a basic computer circuit that could operate at a speed of twenty kilohertz. By the following spring, at the inaugural meeting of the NSA's Working Group on Ultra High-Speed Computer Circuits in Corona, California, Zoltan Bay had already developed a circuit for the NSA that could do the same thing at one hundred megahertz—that is, five thousand times faster.

Buck was only part-time with the NSA. Yet, increasingly he was clearly part of the agency's upper echelons. On occasion he would travel with Lou Tordella, who would go on to become the longest-serving deputy director of the NSA. Almost monthly, Buck would be asked to Washington to brief the Department of Defense or the NSA on the latest developments in the world of computing and how they could be applied to the military.

Bizarrely, in spite of his prolific output Buck was still just a graduate student at MIT. He would have to wait some five years before earning his doctorate. Many of the students who worked alongside Buck at MIT had no idea of his alternative existence as a government agent until they were contacted during the research for this book. To many of them he was just a cheerful young lab rat who ran a scout troop in his spare time and happened to disappear for long periods at a time.

"Dudley Buck has a remarkable ability to apply his knowledge of physics and electrical engineering to engineering problems," wrote his lab supervisor David Brown in a December 1953 staff appraisal—which rated the young scientist as "superior" and recommended a maximum salary hike. "Of all the engineers we have known, he probably derives the most use of his academic training. His understanding of physical principles is excellent. His ideas for new techniques have been very useful for the laboratory program. Another outstanding characteristic is his enthusiasm. This enthusiasm is very stimulating for a group program such as that in the Digital Computer Laboratory. His excellent personality and breadth of interests have made him the recipient of many technical liaison assignments to other laboratories."

With the Soviet threat escalating day by day, Buck's work in the MIT lab and for the NSA were about to move more closely together. Buck had already started to come into contact with Soviet agents, keen to talk about exchanging equipment and information.

IN THE WEEKS after Bar-Hillel's machine translation conference, Buck asked permission of both Dean Acheson, the US secretary of state, and General Walter Bedell Smith, director of the CIA, to visit Moscow.

The details of his request are unclear; the only reference appears in declassified interdepartmental correspondence discussing the idea. It seems that Buck was attempting to procure materials from one of his Soviet counterparts—and possibly to use that as an excuse to find out more about Soviet computing developments.

The creation of the iron curtain had interrupted trade in many basic elements and metals, including many of the key components needed to make experimental computers. It was not always easy to get hold of things that came out of the ground in the wrong countries.

"The DCI [director of central intelligence] instructed me to confirm with Doc [H. Freeman] Matthews of State that the Department had no objection to Mr. Buck visiting Moscow, upon the understanding that no barter deal was contemplated," wrote one anonymous CIA official regarding the request. "I later confirmed this with Undersecretary Matthews, who said the department would accept our judgment as to the value of such a trip, and had no objection thereto, so long as Buck did not undertake to deliver copper or lead, or other prohibited items, to the Soviets. The Department would have no objection to having Buck purchase Manganese for cash."

The USSR had a long-standing company, Amtorg Trading Corporation, dedicated to handling its imports and exports. Amtorg had offices across America, and in the past had signed deals to bring Ford automobiles, International Harvester agricultural machines, and DuPont chemicals to the Soviet Union. It had also brokered the sales of Soviet timber, flax, furs, and caviar.

One way or another, Buck started talking to Amtorg, whose agents seem to have turned up unannounced at MIT to attempt to broker some kind of trade.

As 1952 turned to 1953, the administration of President Harry S. Truman handed power to Dwight D. Eisenhower. The tone of follow-up letters in the new regime suggest that there was a degree of frustration in officialdom about Buck's dealings with the Russians.

Samuel W. Anderson, the assistant secretary of commerce, wrote a memo for Allen Dulles, the new director of the CIA, informing him about Buck's attempted Soviet trades:

Mr. Buck telephoned again to our Mr. Frederick Strauss yesterday and reported that his Amtorg contact had paid him another visit in order to talk further about trade, particularly two-way trade. Mr. Buck stated to Mr. Strauss that he had informed his Amtorg contact that the possibilities of any progress might be substantially advanced if he, Mr. Buck, knew more about what the USSR desired in the way of imports from the US. The Amtorg representative stated that he would attempt to develop this and promised to "cable."

Mr. Strauss has made it clear to Mr. Buck that he is not to disclose his conversations in the Commerce Department. I have asked Mr. Strauss to see to it that I meet Mr. Buck, which is now planned for June 12. Mr. Strauss believes that Mr. Buck may have further information by that time. I have instructed Mr. Strauss to evince my interest in the most casual way.

As you will see from this memorandum, Mr. Strauss, before I knew about this case, has generally informed Mr. Ray Vernon of State, I have instructed him not to continue to advise Mr. Vernon of the situation, unless you wish us to do this. I thought you might have other desires as far as informing the Department of State.

A handwritten scrawl on the archived version of this memo confirms that Dulles saw the note. A second notation indicates that the CIA told Anderson's assistant, Arthur Frazer, that they should keep informing the US Department of State of Buck's movements as they had previously.

Buck traveled extensively during the period between these two letters. He was not only working on the machine translation project, and Blois's experiments with sea mud, but it was also an extremely busy time in the lab.

There is no evidence to show that Buck ever went to Moscow. It is possible that one of the many excursions diarized as a visit to one military installation or another was a cover for a trip behind the iron curtain, yet that seems unlikely.

What it does prove, however, is that Buck's name was known already to two consecutive directors of the CIA—both of whom seem to have considered him an appropriate conduit to trust with back-channel conversations about breaking Cold War trade embargoes.

More to the point, the exchange serves as proof that, seven years before his death, Soviet agents already knew about Dudley Buck—and had a habit of turning up at his office to make demands of his time.

10

TWO SMALL WIRES

AMONG THE MANY SKETCHES THAT LINGERED IN DUDLEY BUCK'S notebook as he flitted between multiple projects was a design for a device made from two small wires, which he called a Bismutron. Bismuth is a brittle white metal, often confused with lead at first sight. It can be made extremely resistant to electricity using magnets; with just a small magnetic force it can block electricity altogether. Buck believed that resistance could be used to create computer "bits." The first iteration of the idea was noted in the spring of 1952, two days before he was flown to Corona, California, to be inducted back into the intelligence community. He started working on it in the lab soon thereafter.

He took two bismuth wires and managed to use them to rig up a basic computer circuit. If he passed a current briefly through the first loop of wire, the circuit could represent a zero. If he did the same to the second, it represented a one. As a pair, the two wires could create a binary digit, and do so quickly. They could show zero or one. One of the best things about it was that it had no moving parts.

The idea worked neatly and was witnessed in Buck's notebook by his lab partner Ken Olsen, who would go on to make a fortune creating Digital Equipment Corporation, one of America's first office computer companies. There was a problem, however. To make the Bismutron work, it had to be hooked up to other computer circuits using the existing technologies of valves and switches. And it needed the small amount of current to keep passing through it in order to maintain the one or zero that had been stored.

The Bismutron was a bit pointless, in other words. For a brief moment, Buck was the laughingstock of the Digital Computing Laboratory at MIT. That did not stop him from mentioning his new gadget to his peers at the NSA, however. For he was confident the idea was a good one.

In between his day job testing magnetic memories for the Whirlwind computer and his extracurricular activities with nuclear mud from the bottom of the Pacific, Buck used his time in the lab to plug away at his idea.

Throughout this time, Jackie hardly saw her fiancé. He was out of town a lot, and when he was back in Boston he was often in the lab for hours at a time. He and Olsen were at their most productive period on the Whirlwind project, designing and patenting computer components that would go on to generate millions of dollars for MIT.

Although he had not yet even been awarded his doctorate by MIT, Buck was being asked to lecture all over the country. Also, through his NSA connections, he had picked up another occasional consulting contract with the RAND Corporation in Santa Monica, California, earning him the princely sum of forty-five dollars a day.

For Jackie, dates with Buck became five-minute sandwich lunches over his workbench in MIT's Digital Computing Laboratory.

The Bismutron was stuck in Buck's head. Although he acknowledged it did not work, he was certain there was a valid idea in there somewhere. Eventually, the cold Massachusetts winter sparked a brain wave.

It was Saturday, December 15, 1953. Buck had taken on the running of the Wilmington Scout Troop, mostly out of interest for his foster son, Glenn Campbell. That morning he had arranged a brisk winter hike out to their campground, five miles north of town. The land had been donated to the scouts by a local attorney and it played host to an annual "camporee" every May. Throughout the year Buck would make the boys hike out to the camp and back fairly regularly.

It appears that on that wintry ten-mile round-trip, Buck came up with a modification to his Bismutron idea. What about making a switch that operated at subzero temperatures?

His experiments with deuterium had taught him a little about cryogenic substances, as had the Ivy Mike hydrogen bomb. At MIT, meanwhile, the physics department had perfected the art of making liquid helium—which was allowing all manner of experiments to be carried out at temperatures as low as -269 degrees Celsius. Temperatures that low opened up the world of superconductors—metals that conduct electricity with no resistance at all if they are cold enough.

If Buck used superconductors for his tiny switches, the current would only need to be applied once for it to maintain a zero or one, solving one of the key problems with the Bismutron. It had also been proven that superconductors stopped conducting electricity instantly once exposed to a magnetic field. It should be possible, therefore, to use that rapid change of state to create an ultrafast switch. Superconducting metals, in their cryogenic state, would register as a one. The interrupting force of the magnetic field could then flip it back to zero.

Buck got into the lab first thing on Monday morning, grabbed his notebook, and scribbled down the idea: "Use hysteresis of superconductors as the basis for a superconducting matrix memory. The superconducting state could represent a ONE and the normal state a ZERO." The note was witnessed by another researcher named John B. Goodenough, a name that rarely popped up in Buck's notes either before or after this time. He seems to have been the only other person in the lab that early who was qualified to declare that Buck's idea had been "discussed and understood."

Superconductors had been known about for forty-two years, but no one had found any practical use for them. They were just one of those interesting physical phenomena discovered almost by accident.

A scientist at Leiden University in the Netherlands named H. Kamerlingh Onnes had worked out how to liquefy helium in about 1908. Three years later he started experimenting on how various metals behaved in this new extremely cold environment he had created. He found that the resistance of mercury suddenly dropped to zero after the temperature reached a certain point. Other metals—such as aluminum and titanium—did the same thing at different temperatures.

Superconductors sounded too good to be true. What they promised was, in essence, perpetual motion: if there was genuinely zero resistance to an electric current it would keep running indefinitely until some other force was applied. The only problem was that the phenomenon only existed at these extremely low temperatures that had to be artificially created by using huge amounts of energy.

S. C. Collins at MIT had built his own machine to liquefy helium immediately after the war, and had taken the experiments to whole new levels. He found that metals that were normally good at conducting electricity—like gold, silver, and copper—were unchanged by exposure to the low temperatures.

Yet some metals that naturally blocked an electric current at room temperature became superconductors once they were steeped in a vat of liquid helium; the most extreme examples were lead and two much rarer metals named tantalum and niobium.

It was this last group of superconductors that Buck was interested in. If he applied a magnetic force at the right temperature, these largely unknown metals could potentially become the backbone for a minuscule, and ultrafast, computer.

Buck knew that to expand on his theory he would have to get hold of wire made from these different superconducting elements and a reasonable amount of liquid helium.

With the university closed for Christmas, he worked day and night on experiments for the Whirlwind project and assorted jobs for the NSA. On New Year's Eve he filed a big report on magnetic computer memories for General Electronic Laboratories in Boston related to a contract they had won to build a computer for the US Bureau of Ships.

As 1954 arrived, he was working furiously on his new idea, leaving the house in Wilmington at the crack of dawn and returning late in the evening. By January 6 he had worked up detailed sketches of how to build and test his new "superconducting switch"—and then start to hook it up to other circuits, assuming it could be made to work.

After a brief trip to New York to give a lecture on his Whirlwind work for the American Institute of Electrical Engineers, Buck cleared his diary to press on with the work.

By February 10, he had successfully begged Collins to give him regular access to his helium tanks, with the agreement he would rent them out hour by hour. He then conceived a test to check that the metals were indeed in a superconducting state inside the helium flask, witnessed by his lab partner Bill Papian. He took further measurements to check the temperature range at which the metals would remain in this state, again witnessed by Papian. Buck also had to build a special metal contraption to lower his gadget into the cryogenic freezer of liquid helium—which spewed plumes of dry ice clouds across the room. By Valentine's Day, he had worked out how to get a strip of tantalum with holes punched in it. The magnetic control wires that would switch it on and off were to be fed through the holes.

To the untrained eye, Buck's device was just one bit of wire with a slightly different colored wire wrapped around it, strapped to the end of a metal pole. By February 18, 1954, after one failed test run, the superconducting switch he had conceived appeared to work. It had been able to switch from conducting a current, to resisting it.

When he returned to the lab on Monday, Buck got back to his new pet project with all guns blazing. He only took a break to have lunch (again, sandwiches in the lab) with Jackie, who had not seen him all weekend; he had been leading the first scout camp of the year.

They chatted over the steaming vat of liquid helium that Buck was using for his experiments. He explained to his future wife in loose terms what it was he had found.

"What are you going to call it?" she asked.

He confessed that the term "superconducting switch" lacked a certain lyrical quality. He was toying with the idea of calling it a "cryistor"—smashing together the words "cryogenic" and "transistor." It was literal enough for the scientists to get it, and yet had a futuristic ring to it.

Jackie wasn't so sure. The word looked a bit ugly. "Why don't you call it the cryotron?"

11

THE CRYOTRON

O N JUNE 6, 1954, DUDLEY BUCK AND JACKIE WRAY WALKED DOWN the aisle at Unity Church in her hometown of North Easton, Massachusetts. They had set the date one week after Jackie named the cryotron, over a dinner with Otis Maxfield, the minister at the Wilmington Methodist Church, whom they wanted to conduct the service.

Schorry Schick, Buck's lodger and future brother-in-law, was his best man. They had grown close in their months together, in spite of tensions from the neighbors—some of whom were still a bit paranoid about a German living in their midst in the years following World War II.

Jackie's sister Gwen was maid of honor and Nancy Whitman, the mutual friend who had introduced the young lovers, was bridesmaid. George Frideric Handel's "Largo" played throughout the service in the pretty little church.

Glenn Campbell was a notable absentee. He had gone back to Washington, DC, shortly before the service to stay with his brother, fearful that he would get in the way of the newlyweds. No matter how much they protested, Glenn refused to listen. He grabbed his bags, and the $150 Tosi accordion that Buck had just bought him, and caught a bus home.

It wasn't just that he wanted to get out of their hair. Letters from his brother Bill were filled with tales of parties and dancing, which sounded a lot more fun than anything Wilmington had to offer. Yet Glenn never lost touch with his foster father, and to his dying day kept the last letter he ever received from Buck in the drawer of his bedside table.

Buck had been making preparations for married life. The Plymouth sports car was on its last legs: every other week it conked out on him. He would fix it himself using parts he found in Goldie's Junk Yard in a town nearby. He rigged a chain to one of the sturdy pine trees by the driveway that he would use to lift the engine block off its mount and lower it back again.

As his wedding day approached, Buck splashed out $645 on a maroon 1950 Studebaker Champion, a bullet-nosed two-door sports car. Straight after the wedding reception, Mr. and Mrs. Buck jumped into the car and set off on the long drive to Santa Barbara so the young scientist could introduce his wife to his father, his Grandma Delia, his brother Frank, and the rest of the Buck family.

Although it was their honeymoon, Buck was never one to let an opportunity go to waste. They stopped off in Washington on their way across the country so Buck could pick up some instructions from one of his NSA handlers to carry with him to Magnavox, one of the agency's West Coast contractors. Before the newlyweds hit the road again, the NSA gave him his wedding present: official orders to go back to the National Bureau of Standards lab in Corona, California, between the dates of June 9 and 21. Buck does not seem to have visited the Bureau of Standards computer lab at all on this trip, but he certainly claimed back his gasoline money for the whole round trip to California as an expense. Thus, American taxpayers picked up the bill for the Bucks' honeymoon.

The Buck newlyweds drove through Virginia along Skyline Drive that runs through the Blue Ridge mountains, then through Tennessee, Georgia, Alabama, and on to New Orleans and then Houston before trekking west to Santa Barbara. The journey took about a week.

Jackie felt instantly at home in Grandma Delia's expansive bungalow. She was astonished how big it was inside relative to its humble exterior. Compared to her own "Waspy" New England family, Jackie found the warmth of the aunts, uncles, and cousins a breath of fresh air. There was an ease to their laughter she had never encountered before. And the entire family was impressed by her in turn.

After a few days of laughing, joking, and eating, the newlyweds turned around and headed for home again. By the time they got back to Wilmington, the house was empty. Not only had Glenn gone, but now Schorry had left too. Buck's sister Virginia had been sent back from Berlin to a new posting in Washington; Schorry had packed up as soon as he got the news, and rushed south to meet her plane.

They gave the place a lick of paint and stripped down and repainted an old China cabinet that was being thrown out by one of the neighbors. Buck built a bookcase along one of the living room walls. The sofa and armchair were reupholstered, and proper curtains were fitted to every window for the first time. It was a happy, cheerful place where they started to get to know one another properly. The two young lovers had only been together for sixteen months before they got married.

Once they had the inside of the house looking up to scratch, Buck set to work on the garden, begging some rich, loamy soil from a local dairy farmer, and then planting a lawn, followed by some yew trees, laurel bushes, and rhododendrons that he neatly placed around the edge of the house.

He was preparing their home for the family they both wanted. In the lab, meanwhile, Buck's other baby was taking shape.

AT THE HEART of the sprawling campus of MIT sits Building 10—the pillar-fronted palatial construct topped with its famous "great dome" modeled on the Pantheon in Rome.

On the third floor of this imposing edifice, Buck—and a handful of other young electrical engineers and physicists who were working on different computer technologies—were given some lab space.

Wires and cables flew everywhere around the room, firing up strange whirring machines on the workbenches. Buck's experiments, conducted at temperatures close to absolute zero, stood out from the crowd.

Liquid helium would arrive from the physics department in giant dewar flasks—the laboratory equivalent of thermoses. Buck was trying to find the right combination of metals to make his cryotron work. He was taking delivery of wires made from chemicals that most of the staff had only seen before as symbols on the periodic table of elements.

Initially he had tried to make the device using the cheapest super-conductor: lead. He struggled to find lead that was pure enough to perform the task, however. The thickness of the wire was also a consideration. He ordered tantalum wire of 0.007" thickness, and 0.003" niobium wire in one-thousand-foot rolls. At room tempera-ture, wires that are thinner and longer have higher resistance. It turned out that this was not the case at cryogenic temperatures: a supercon-ductor was a superconductor.

After learning of his experiments, General Electric agreed to donate a small sample of an even rarer superconductor called rhenium. Only a few years earlier, rhenium had sold for about ten thousand dollars a gram, but it had become more freely available after World War II.

Buck's work was creating a stir, and the word cryotron had en-tered the lexicon of Building 10. Within weeks of Jackie coining the phrase, there were master's-degree students choosing to write their theses on the tiny gadget.

Buck had started to explain to Howard Campaigne, Joe Eachus, and Solomon Kullback at the NSA how his cryotron could hold the key to faster, smaller computers. If it could be perfected, the new de-vice could be applied to almost all of the problems they were trying to solve in terms of codebreaking, large-scale data processing, and possibly more aggressive military applications.

Still, it was not much to look at. The cryotron was one straight wire, about one inch long, with a second wire wrapped around it. Both wires were made from superconductors but using different metals that hit their magical state of superconductivity at different temperatures, which added a little to the complexity of the early experiments.

By passing a current through the second wire Buck created a mag-netic field that would see the straight wire running through the middle switch from being a superconductor to a resistor. Once it did that, he could create a computer circuit: the cryotron could flip from on to off, yes or no, one or zero.

There were a lot of variables to contend with before he could get there. He had to test the reliability of the two metals that would com-prise the device. He had to make sure that the temperature in the flask did not get affected by heat from some of the other apparatus needed to make it all work. There was no textbook to work from or improve upon; it was a new field of experimentation. Trial and error was the only way to get to the right answers.

A miniature production line started. The lab secretaries, who had grown accustomed to being asked to thread little magnets onto wires to make parts for the Whirlwind machine, now found themselves wrapping tiny pieces of wire around other tiny pieces of wire. After a few weeks, Buck built a little spindle machine to wind the cryotrons for him in an attempt to bring more consistency to his experiments and to spare the secretaries' grumbles.

By November 1954 the twenty-first incarnation of the cryotron appeared to work properly and consistently as a switch. One wire was made of tantalum, the other niobium.

The very next week Buck went to Washington to tell Eachus and the rest of his NSA handlers all about his new device. Things started to move quite rapidly. The cryotron was a long way from perfect, but Buck was confident that he understood its shortcomings and had credible thoughts on how to tackle them.

Academically, Buck's career was also starting to progress. In January 1955, two months after he got the first cryotron prototype to work, he finally got his first teaching job, giving classes on computer control components at Northeastern University's evening school. He still worked at MIT by day, and lectured at Northeastern on a handful of nights a week.

He then got another job offer, this time from Arthur D. Little, the chemical analytics company (which has since evolved into a management consultancy). Buck was offered a retainer of two hundred dollars a month. His task? To make more cryotrons. Little, which was closely tied to numerous government research projects, wanted to pursue Buck's idea commercially. They suggested he work for them on a part-time basis, to make sure they were on the right track.

It was an easy offer to accept. The company had one thing in abundance that Buck sorely needed: liquid helium. Little had built the first commercially available helium liquefiers, but at a princely sum of twenty-three thousand dollars such a device was somewhat beyond his research budget.

The helium liquefier in the MIT physics lab could produce twenty-seven liters of liquid helium an hour. Arthur D. Little's machines could churn out only four liters an hour, but they had dozens of them. Working with Little could see Buck get his hands on sufficient quantities of liquid helium to experiment on multiple ideas simultaneously.

Helium gas supplies had started to pose problems for Buck by this point. S. C. Collins at MIT had been receiving stocks of helium for free, directly from the US Department of Defense, for his initial experiments, but Buck's substantial requirements had put a stop to this pro bono support of science. Collins had been instructed that any future supplies of helium gas be ordered with the Boston Naval Shipyard at a price of $4.20 a cylinder.

Buck's new arrangement with Arthur D. Little was fortuitously timed. Although he had yet to announce the cryotron publicly to the world, the rival scientists operating as part of the NSA's industrial cooperation program were well appraised of his progress. He was happy to explain the concept of the cryotron and the broad outline of his work to almost anyone who enquired about it, whether they were commercial rivals or academics at other institutions.

Although a flood of new commercial consulting offers appeared, Buck started turning down all new approaches to work on the cryotron. In one rejection letter to a prospective employer, Buck explained the progress he was making with the device he had built in the Arthur D. Little lab. "The work is going along smoothly," he wrote. "We have three small flip-flop clock circuits which operate at about 1 kilocycle per second and are trying to work with finer wires and a variety of materials and temperatures in order to push the speed up toward the megacycle region. We are being forced to learn a lot about metallurgy in dealing with fine wires. The prospects for a major breakthrough in this computer field with this new component look as good as when I visited you."

With a queue of America's new corporate titans desperate to hire Buck, MIT eventually gave him a teaching job in July 1955, appointing him as an instructor in electrical engineering on a salary of $3,420 a year, which worked out at about an extra seventy dollars per month.

A fortnight later he turned down a job offer from the industrial conglomerate Westinghouse, which offered to pay him a colossal starting salary of $8,700 a year. It was an "attractive offer" for a job "doing almost exactly what I hope eventually to do," Buck wrote back. "But I have weighed the relative advantages of the two possibilities and I feel it is to my advantage to invest another year in academic work at MIT. I am deeply grateful to you for your time and interest in interviewing me."

Buck had committed to MIT, and MIT had committed to him. Now that he was finally a proper member of the department, he published a paper on the cryotron to circulate among the teaching staff. The abstract of the paper spelled out clearly what he had just created:

> The study of nonlinearities in nature suitable for computer use has led to the cryotron, a device based on the destruction of superconductivity by a magnetic field. The cryotron, in its simplest form, consists of a straight piece of wire about one inch long with a single-layer control winding wound over it. Current in the control winding creates a magnetic field which causes the central wire to change from its superconducting state to its normal state. The device has current gain, that is, a small current can control a larger current and it has power gain so that cryotrons can be interconnected in logical networks as active elements. The device is also small, light, easily fabricated, and dissipates very little power.

Superconductors were considered highly experimental. There was a great deal of skepticism about the claim that superconduction equated to perpetual motion. Before even spelling out the nature of his own invention, Buck felt compelled to lay out a defense of the broader field of cryogenic research: "Below the superconductive transition the resistivity is exactly zero. That it is truly zero is vividly demonstrated by an experiment now in progress by Professor S. C. Collins at M.I.T. wherein a lead ring has been carrying an induced current of several hundred amperes since March 16, 1954 without any observable change in the magnitude of the current."

The paper was dated August 22, 1955, so Collins's experiment had been running uninterrupted for seventeen months. Resolving that he had laid that doubt to rest, Buck proceeded to explain how he had been able to wire large groups of cryotrons together.

The device worked perfectly as a computer memory, Buck explained, and could also be used to control other types of computer memory and work with filing systems to code and decode where data had been stored. It could add, subtract, multiply, and divide. It could be used in electronic logic circuits; it could operate as an amplifier or with a power control circuit; it could be used to convert analog data received from real world inputs into a digital signal, and then convert it back again.

In different groupings and configurations, these two little pieces of wire suspended in helium could be used to create every type of circuit needed to build a computer. While most computers still occupied whole buildings, this one would fit in a box that was just one foot square.

> The cryotron in its present state of development is a new circuit component having power gain and current gain so that it can be used as an active element in logical circuits. It is easily and inexpensively fabricated from commercially available materials and its size is small. Extrapolating the volume occupied by the present experimental circuits to larger numbers of components indicates that a large-scale digital computer can be made to occupy one cubic foot, exclusive of refrigeration and terminal equipment. The power required by such a machine extrapolates to about one-half watt, once again excluding refrigeration and terminal equipment. The reliability of cryotron circuitry is not known, but it is anticipated that operation in an inert helium atmosphere at a temperature near to absolute zero where chemical activity and diffusion processes are essentially stopped promises a high degree of reliability. The circuit noise level is similarly not known, but due to the low temperature, very little thermal fluctuation noise is anticipated. The device is at present somewhat faster than electromechanical relays, but far slower than vacuum tubes and transistors. A program is under way to increase the speed.

Dudley Buck had invented a whole new field of physics and electrical engineering. Though his name disappeared from public view after his death, his legacy runs deep through the veins of the NSA, and computer science, to this day.

As David Brock from the Computer History Museum in Mountain View, California, explains,

> Is there someplace else where some person may have made a superconducting switch? Probably. But Dudley Buck is the person who got this whole field of superconducting electronics going.
>
> Since Dudley Buck's time, it has been a constant dream of the NSA to build its frozen supercomputer. The NSA has never given up on the promise. If you can make a computer that uses almost no power, you can have these gigantic systems. The NSA needs all this capacious memory. They need the sheer processing power. You can see the problems they get into with things like this data center they built in Utah—the power in that thing is unbelievable. They have had fires in there. Power limitations are very real.
>
> The promise of superconducting electronics, where it almost uses close to no energy per unit of computation, means they don't need to care.
>
> They have tried consistently. What they have failed to do is to invest, and to persuade industry to invest. But they try, to this day. There's a multi, tens-of-millions of dollars project now being led by IARPA [Intelligence Advanced Research Projects Activity], the intelligence community's research and development agency, to build a superconducting computer. It's built of cryotrons. They call them something else—tunneling Josephson junctions. But those are cryotrons.

BUCK HAD BEEN married for a little over a year by the time the cryotron was starting to create a fuss. In the summer of 1955, as interest in the device reached fever pitch, he had a different set of pressures to deal with. At home in Wilmington, Jackie was pregnant with their first child.

Having first mentioned children on their third date, Buck was ecstatic about the impending arrival. He would take home T-bone steaks and insist that Jackie eat the tenderloin. "Growing babies need good building blocks," he repeatedly reminded her, "quality amino acids, protein."

The summer of 1955 was a brutally hot one in Massachusetts. Jackie, weighed down with the baby, was struggling to cope. Buck, in an attempt to distract Jackie, came up with a bizarre idea.

He took home a soldering iron and a Heathkit amplifier that would become the center of a new sound system for their house. Jackie, who had never made any pretense at being technically minded, soon found herself being taught how to make perfect electrical joint connections. The amplifier she built while eight months pregnant worked flawlessly for years to come.

When the baby was due, Buck refused to leave the hospital until the baby had been born safely, subsisting on packets of M&Ms from the vending machine. The idea of a father coming into the delivery room was still very much taboo, but Buck hovered by the door for hours.

Eventually, at 7:42 p.m. on September 4, 1955, Carolyn Buck was born, weighing nine pounds and five ounces. Buck could hardly contain himself. In the days before the baby was allowed home, he bounced in and out of the hospital three times a day: morning, noon, and night. He would gaze adoringly through the nursery window. The nurses were so charmed by his doting behavior that they let him stay long after visiting hours were over.

On the day they took Carolyn home, Jackie's mother and sister Gwen were waiting for them in the doorway. Buck took the baby from her mother's arms and held her up for the waiting neighbors to see.

"It's a boy next year," he proclaimed.

His mother-in-law's jaw dropped. "Oh my God, did you hear what he said?"

12

LAB RATS

CHUCK CRAWFORD WALKED INTO BUILDING 10 AT MIT, HIS STOMACH knotted with nerves. He stared up toward the impressive domed ceiling and the four stories of balconies that edged the cavernous entrance hall. On a podium in front of him there was a map of the whole campus to help strangers navigate their way around—the complex numerical naming system that coded every room in each of the buildings often left visitors confused. The map had rows of buttons that lit up the route to each building when you pressed them.

There was also a red telephone, and an internal directory listing the names and numbers of all the professors and tutors. Crawford looked up the number for Dudley Buck, picked up the receiver, then hesitated. Could he make the call? Should he?

The autumn term of 1956 had just begun. Crawford was now in his second year of his physics course. A lecture he had heard before the summer break was still running through his head. Dudley Buck had explained how computers would change the world. He spelled out the need to make computers smaller, so they could in turn become quicker and more energy efficient. He spoke about how components had to be designed in a way that could see them mass-produced. He told the freshman students a little about his cryotron.

"I listened to that lecture and I thought this young professor really knows what's going on," remembers Crawford. "He was ahead of his time. I thought about it all summer, that one-hour lecture. By the time I came back in the fall of '56, well, I guess you have to call me driven. Foolish would be another word."

Crawford had spent his summer working in a computer factory, assembling machines that were using the vacuum tube technology. He worked hard, putting in as many hours as possible. He needed the money—being from a nice, middle-class background meant that Crawford had to pay most of his tuition himself. Schemes such as the V-12 program that had put Buck through his education were no longer in existence.

During the semester, Crawford had a job in the dining service at MIT. "It was a pretty terrible job, but it was a job," he says. "Without that job, I wouldn't have had enough money to eat."

He had come back to campus determined to dedicate his time to getting close to Buck, even if it meant he would have to go hungry.

"Dudley's phone number was 622," Crawford remembers. "It became my telephone later. I picked up the phone there, in the main entrance lobby. It was extremely intimidating for a fledgling sophomore. I figured I would get a secretary that maybe I could talk into making an appointment for me to see Buck. I dialed 622 and waited. The person on the other end of the line said one word: 'Buck!'"

"Professor Buck, how would you like a student technician to work for you for free?" asked Crawford, with a tremor in his voice—forgetting that Buck did not even have his doctorate yet, never mind a professorship.

"For free?" asked Buck, incredulously, before pausing for a few moments. "Come right up!"

Buck's office was the first in a row of a half dozen with windows that looked out on a shared workspace, where oscilloscopes, microscopes, and electric wiring weaved around the vials of helium tied to his cryotron experiments. His office was the regimented space one would expect of a military man, furnished with a plain wooden desk and a wooden swivel chair. Buck still had the perfectly straight back and upright posture he had learned during his navy training in Seattle.

After climbing the stairs, Crawford knocked on the lab door. The first thing that struck him was the steady, throbbing hum of the dozens of vacuum pumps attached to experiments around the room. The young student barely had time to introduce himself before Buck sat him down at the bench immediately outside his office door and gave him his first problem to solve.

"He handed me his wristwatch, which didn't work," remembers Crawford. "I opened it up, looked at it with a microscope, got a pair of tweezers and figured out what was wrong with it—which wasn't much, it was just some dirt in it or something. I fixed his wristwatch for him, so he gave me other things to do."

Like Buck, Chuck Crawford had experimented a lot as a teenager. Crawford's mother had always wanted him to go to MIT. She was the youngest of eight children and her eldest (and favorite) brother, who was eighteen years her senior, had studied there and then became a mining engineer in British Columbia. He died during the Spanish flu epidemic of 1918.

Before Crawford was even born, she had always wanted him to follow in his uncle's footsteps. They had moved to South Orange, New Jersey, as she had heard there was a good high school there. She cleared a space in the basement so that young Crawford had a place to do experiments.

Crawford thought at first that he might want to be a horse trainer, but soon got the bug for electronics. Over time he built his own oscilloscope and countless radios. He won a nationwide competition for school students by making a working computer with electrical relays he had coaxed out of AT&T, the telephone company where his father worked.

He almost blew up his parents' house when he decided to build a pulse jet engine in the garage. He was also accustomed to working with tiny parts, having built model ships and airplanes. As a grade-school student he also made jewelry and sold it around town.

By fixing Buck's watch in just a few minutes, Crawford had passed his audition. He was told he could come into the lab ten hours a week. Crawford was delighted, but the time commitment was difficult. It meant he would have to give up the job in the MIT dining room, leaving him short of cash, but Crawford felt it was worth the gamble.

Shortly after his arrival Crawford was tasked with setting up experiments for the cryotron. About a week after he first walked through the door, he was hunched under a lamp, winding one tiny wire around another, when a heavy hand thumped down on his shoulder.

"You!" said Buck, as he leaned into the young student's ear. "You, we're gonna pay."

The ten hours a week went up to twenty—working hours that all came in addition to his study time.

"Suddenly I had a job," Crawford remembers. "Dudley proceeded to pay me more than I could get in the dining service job. Then there were a series of raises. I was in seventh heaven. I just about worshipped the man."

No one else in the lab had as much experience as Crawford in working with tiny tools and manipulating miniscule parts. As a result, he was perfect for the job of making cryotrons. "I suddenly became Dudley's principal technician."

Buck's lab was an interesting melting pot; students from all over the world passed through the door, including many from Eastern Europe who had come to America in search of a new life. A six-foot-tall, blue-eyed Ukrainian grad student once got in a fight after someone accidentally called him Russian. There were others from Austria, Germany, and India, and the occasional Brit. In total there were about 650 foreign students at MIT at the time. About 65 percent of the student population was studying engineering or science.

It was a "work hard, play hard" type of place, but with its own distinctive quirks. A young secretary named Carol Schupbach who worked for Buck and the other young assistant professors occupied a vast room that doubled as an unofficial café for the whole building. The room had previously housed one of the early wartime computers, which had subsequently been moved out to the Barta Building with the Whirlwind machine. Although the computing equipment had been removed, the air conditioning that had been installed to keep the temperature down had never been moved.

Schupbach's office was, quite literally, a cool place to hang out. To try to maintain some privacy from the steady stream of visitors popping in for coffee, the young secretary boxed herself into one of the corners behind a protective wall of bookcases.

"We had a loudspeaker in our lab and we listened to music," remembers Schupbach. "A lot of labs had radios, but they listened to longhair music. We listened to show tunes. Sometimes if the music

THE CRYOTRON FILES | 127

was too loud, Dudley would just close the door. Once I didn't realize he was in his office. The music was on, and the song was "We've got trouble right here in River City," from *The Music Man*. I walked by Dudley's office and seeing him in there, I said, 'Oh, Dudley, I am so sorry, I didn't realize you were here. Is this music bothering you?' In rhythm to the music he replied, 'Not at all, not at all.'"

There was, arguably, still a childish immaturity to Buck that perhaps undermined his credibility with other, older members of the faculty. Yet former students attest that he sparked fierce loyalty among those under his charge. He would often dispense advice in cryptic, folksy sayings. "We live mostly on cans of beans, and the dog licks the cans," was one of his favorites. The unofficial slogan of MIT was "Tech is hell." A number of Buck's alumni confess that he had to talk them out of quitting at various points in their academic careers.

Buck still loved a practical joke, just as he did as a boy. In spite of his increased stature on campus, he was happy to play along with juvenile pranks at any opportunity. One Halloween, Carol Schupbach brought in a jack-o'-lantern. When she came back from lunch, it had been turned into the head of a dancing scarecrow, wearing a lab coat and a hat. A cigar stuck out of the pumpkin's mouth, with a flashing red diode in the end to make it look like it was lit.

On another occasion, she came back to her desk to find a large frog sitting in the brandy glass on her desk that she used to hold pens. No one could tell her why, or where the frog came from. They then decided to keep a pet snake in the lab, called Snively.

One student came home from a weekend away to find that his Volkswagen Beetle had been stripped apart, and then reassembled piece by piece, in his dorm room.

George, the building janitor, was a regular victim of pranks. He liked to tell everyone that he was a union man, citing this as a reason why he didn't have to work too hard. As a result, most of the staff treated him poorly. Buck made a point of being civil to him, for the most part.

Every day at about 4:00 p.m., George would drop by, pushing his trash cart. He was always trying to win the competitions in the Boston Globe and would often ask Buck for help with the clues. "Dudley knew who George was and what George was—he was a goof-off," recalls Crawford. "But he was a human being, and that's all that mattered to Dudley."

Buck had ordered a sample of superglue, which had only just been developed in the Eastman Kodak lab across town. Nobody really knew how powerful it was. On the afternoon that his sample arrived, Buck had just started to open the tube and have a sniff of the transparent compound when George walked in, looking for help with his newspaper competition.

"Hey George, get over here," Buck said. "Stick out your hand," Buck ordered, then placed a tiny drop of the glue on George's index finger. "Now do this," he said, pressing his index finger to his thumb to make an "okay" signal. George did exactly as he was told. Buck then grabbed his hand and squeezed the two fingers a little more tightly together.

"Now pull the fingers apart."

George strained as hard as he could, but his fingers were bonded tight. Buck then grabbed his hand and tried to wrench the fingers apart, but to no avail. A crowd had started to gather around, having heard Buck barking orders. One of the graduate students picked up the instruction sheet that had fallen out when Buck had ripped open the cardboard box with the tube inside.

He started reading aloud: "Be extremely careful not to get this glue on the skin as it is extremely tenacious."

"That's fine, we'll get a solvent," said Buck to George, who had started to panic.

"There is no known solvent," the student replied, still reading from the instruction sheet.

The crowd started to fall about with laughter, as Buck realized he had gone too far.

"I can't work, I need compensation," screamed George, gesticulating his hobbled hand wildly at Buck. The fact that his fingers were still forming an "okay" sign made it all the more hilarious for those watching.

"Give me some of that, I want to get glued to the secretary," howled one of the graduate students, referring to Carol Schupbach.

"Why don't we put a drop on a toilet seat?" chirped another.

Buck took George's thumb, and one of the students his forefinger. The bond between the two was about eight millimeters wide. They teased and pulled until eventually the bond started to break. The layers of dirt on George's hands made it easier to pull them apart without ripping off his skin.

Meanwhile, one of the other students had started to sand down one side of a quarter dollar coin. By the time George's fingers had been liberated, he had filed it dead flat and smooth. He then superglued the coin to the floor of the marble corridor, right outside the bursar's office at ground level. They then watched from the third floor balcony as dozens of people stooped to pick it up.

"Something like two days later, the coin disappeared, but there were chisel marks on the floor," remembers Crawford.

Buck's students never dared to make him the target of their jokes; they knew he would get the better of them in the end. His pranks could border on nasty.

One of his students needed a small amount of platinum for an experiment. As it was an expensive metal, there was a requisition form to fill in that had to be signed by Buck. The order was approved, and the tiny piece of platinum duly arrived.

Buck took the packing slip and added some zeros to the amount that had been ordered. He then found a large roll of stainless steel wire wound into a big coil about fifteen inches in diameter and attached the packing slip. Stainless steel and platinum are both shiny, gray metals. At first glance, they looked pretty similar.

"MIT has its own police force on campus," explains Crawford. "Dudley went downstairs and borrowed a police officer. The graduate student was a very meek guy. They had this fake packing slip, which suggested there was maybe about $100,000 of platinum here. The police officer brought in this big hank of wire, with the slip attached to it. The poor guy got this, and he thought it was his mistake and he'd ordered one hundred times too much. He turned into a whimpering mess until someone pointed out that was stainless steel."

Buck's lab was also used for parties. At the back of the room, disguised as just another innocent experiment, was a miniature distillery that turned out a pleasant enough homemade gin. Buck himself never touched it; he remained resolutely a teetotaler, in line with Grandma Delia's teachings.

The gin inspired some late-night ideas, however. After one party in the lab, a drunken Hungarian student decided to go to the nearest airfield to find a plane that he and a friend could take for a quick spin. It would be a chance to dust off his wartime pilot's training, he argued. They stole a plane, returned it, and apparently no one knew a thing.

Another impromptu late-night drinking session saw a small group of students perform a weather-related experiment with some kites that ended up crashing into power lines and knocking out half the Boston electricity grid.

Buck and the rest of the teaching staff seem to have turned a blind eye to the hijinks. The team in the lab came from very different backgrounds: men who had seen active service during World War II mixed with fresh-faced undergrads, while others were newly back from the front line in Korea.

Not everyone was a student. There were also a handful of engineers and technicians who worked in the lab full-time without any teaching responsibilities or academic agenda.

The most talented by far was Ken Shoulders, considered by those working alongside him to be one of the brightest engineers at MIT. Born and raised in Texas, he was a prickly character who in his brief career had already worked for Collins Radio Company, Continental Electronics, National Geophysical, Radioplane, and Texas Instruments before he arrived at MIT. His experience ranged from microwave radio links to miniature vacuum tubes. He seemed to have no interest in collecting degrees, unlike the academics around him. Shoulders and Buck came to work as a close-knit team.

"Ken was kind of a dreamer," explains Crawford, who now runs his own high-tech engineering firm in New Hampshire and has become one of America's leading antismoking campaigners. "Ken was an outsider, academically—Dudley had to protect him. The set of people that all had PhDs didn't want to recognize that someone who didn't have a doctorate could be as bright as Ken Shoulders was. Ken provided a lot of the ideas that Dudley grabbed on to. Ken would take any idea and try to prove it to its ultimate limit. Dudley was more practical and would say, 'Well, look, we could build these things in a year from now.' Dudley could get things done."

SHORTLY AFTER HE joined the MIT teaching staff in the summer of 1955, Buck was invited to present a paper on his cryotron to the American Institute of Electrical Engineers (AIEE). It told the same story as the document he had prepared for his colleagues at MIT: that the cryotron—two small wires suspended in liquid helium—could replace vacuum tubes inside computers.

The publication of his paper on the cryotron had piqued interest across America.

IBM seemed desperate to hire Buck. Some executives at the nascent computing giant would have been familiar with Buck's name from patents filed on the magnetic core memories that were being deployed in its machines. Within a few months of Buck's cryotron paper being published, however, there were several senior executives from the company's headquarters in Poughkeepsie, New York, who were in regular correspondence with him.

John C. McPherson, IBM's vice president, wrote to Buck shortly after he delivered his presentation to the AIEE. McPherson informed Buck that he had held conversations with William F. Friedman, the head cryptographer at the NSA, about Buck's "superconducting computer components." It appears to have been part of an attempt to brag about IBM's connections; an attempt to impress the young scientist. McPherson clearly had no idea that Buck had developed the cryotron in close association with the NSA all along. Evidently Friedman, who was housebound, recovering from a heart attack at the time, did not tell the IBM man about Buck's connections either.

Yet Buck was always happy to talk about his work. He maintained correspondence with several IBM executives, and made regular trips to its Poughkeepsie headquarters to demonstrate the device. Soon, IBM was running its own experiments with cryotrons.

"I would certainly like to thank you and your associates for the wonderful co-operation you have given us in conjunction with the cryotron work," wrote IBM's Donald R. Young to Buck in February 1956. "The visit with you was stimulating and instructive and would certainly like to encourage you to visit us at any time that you think it might be profitable for you."

Young came to the Buck home in Wilmington that April, in a further effort to persuade the family that Buck should leave academia. In a letter sent after his trip, Young wrote that IBM "had agreed on a very attractive offer and I feel certain that we can provide you with the working conditions you desire." Buck told him, as he had every other corporate suitor, that he wanted to finish his doctorate first.

Cryotron projects were starting to show promise. By the time IBM's interest had been fully aroused, Arthur D. Little had already won a contract from the NSA to build a whole computer using cryotrons, with Buck continuing to advise on the project in his role as a consultant to the firm.

"The primary aim is to construct a word-recognition device of sufficient storage capacity to demonstrate operational feasibility of the cryotron and large assemblies of cryotrons," wrote A.D. Little executive Howard McMahon to Al Slade at the NSA. He outlined the design they were using, sending a carbon copy to Buck that remains in his files. "Needless to say, we would hope to use the ideas and direct help of Mr. Dudley A. Buck to whatever extent they are available."

MIT had started to wake up to the commercial potential of Dudley Buck's invention. The institute had struck a deal with a company called Research Corporation in New York that took promising patents from MIT laboratories and then tried to sell them to big businesses—in exchange for a cut of the profits. Research Corporation had started to display an interest in the cryotron, upon the advice of the MIT patent committee.

On February 6, 1956, the MIT news service felt compelled to issue a press release for the morning papers on the "pioneering device to replace transistors and tubes in giant computers" that paved the way for what Buck had told them would be the "coming revolution in electronics":

The new device is a cryotron, so small that 100 will fit into a thimble. It is the first useful application of a phenomenon discovered nearly 50 years ago but still not yet understood.

Development of the cryotron was begun three years ago by Dudley A. Buck, a graduate student and instructor in the Electrical Engineering department at MIT, in conjunction with the Lincoln Laboratory.

The first data-processing equipment in which this simple, tiny device will replace complex tubes and expensive transistors is now being built at Arthur D. Little, Inc., with the cooperation of MIT engineers.

This first cryotron electronic catalogue will use 215,000 cryotrons. A conventional computer to do the same job might require more than 50,000 vacuum tubes.

Present experimental circuits, says Mr. Buck, suggest "that a large-scale digital computer can be made to occupy one cubic foot", not including refrigeration and terminal equipment. In contrast, today's digital computers fill whole rooms.

There was also interest from military contractors. A delegation from Glenn L. Martin Corporation, the famous defense contractor now part of Lockheed Martin, came to visit Buck's lab for a full demonstration of the cryotron research program.

The company had been making military aircraft since World War I, and had built 531 Boeing B-29 Superfortresses during World War II, including the two planes that dropped the bombs on Japan. It was now starting to move into the postwar world of missiles, which entailed more automation and computerization. Hence, the cryotron was of interest.

The US Air Force also took an interest in Buck and his inventions. Colonel Otto G. Quanrud, director of control and guidance for the air force, based at Griffiss Air Force Base in Rome, New York, got in touch after hearing his presentation to the AIEE. He wanted Buck to build a "cryogenic gyroscope."

"Among the problems presented by Rome Air Development Center, the one pertaining to the development of a cryogenic gyroscope is considered to have the greatest potential impact in the fields of navigational traffic control and missile guidance," Quanrud wrote. "The concept of the utilization of the rotational moment of inertia of a current in a superconducting medium is not new, but the value of the small, light-weight, very low drift gyroscope which would result in the event that such a development were successful would be so great it is felt that expenditure of high-level, scientific manpower resources on such a project is warranted. As you are noted for your work in low temperature physics, it is considered that this problem might be of interest to your institution."

Colonel Quanrud also appears to have been unaware of Buck's role with the NSA, addressing him solely in his capacity as a researcher in MIT's computer components laboratory. As a navy man through and through, it would have been unseemly for Buck to cooperate with the brash new arm of the American military that was the US Air Force. Buck suggested that the colonel get in touch with the Arthur D. Little Company, which might be able to help him. He did not explain that he was employed as a consultant by the firm.

A handful of MIT alumni who had worked with Buck were now starting to make their way in the commercial world, and spreading the gospel of the cryotron too. Gordon Burrer, a student of Buck's who had written his thesis on the cryotron, had managed to spark interest from his employers at Avco Manufacturing Corporation in Cincinnati. Burrer wrote to Buck asking for everything that was publicly available about the project.

"Cryotron research is in full swing, with seven theses in progress," Buck replied. "We now have our own helium can in Building 10, plus a spotwelder, cryotron winder etc. Let us know what you are doing at AVCO!"

Other students tied to cryotron research were being lured away by big names in electronics. One was poached by William Shockley, the Nobel Prize–winning co-inventor of the transistor, who was working with Beckman Instruments in Palo Alto, California. "We are not anxious to lose Mr. Konkle, and will make him a substantial offer ourselves through Lincoln Laboratory," wrote Buck in a grudging recommendation letter. "But as an academic institution, we must not become as a librarian who does not like to see her books circulate."

Shockley developed an interest in the cryotron and came to MIT for lunch with Buck. Word continued to spread, in spite of continued negativity about superconductors and lingering questions about the equipment required to create and maintain the cryogenic state in which Buck's invention had to operate.

In April 1956 Buck was asked to write another paper on the cryotron for publication by the Institute of Radio Engineers. The content was little changed from the original internal briefing note he had circulated around MIT the previous summer.

Yet the tone was at times defensive, reflecting the litany of questions about the practicalities of conducting experiments at temperatures close to absolute zero, with equipment that had to be doused in a bath of liquid helium for hours at a time.

"Many materials are used in the construction of circuits to operate in liquid helium," Buck wrote. "Ordinary wire insulation (enamel, silk, glass, Formex, Formvar etc.) shows no sign of failure after repeated immersion. One experiment using wooded coil forms glued together with Duco cement was successful, Scotch electrical tape, while it freezes, seems to hold well. . . . Metals in general are much stronger at extremely low temperatures."

Buck's paper concluded that the cryotron's switching speed was still too slow relative to its theoretical potential, adding that "a program is underway to increase the speed." Rather than using wire, Buck wrote, he was investigating whether he could use thin layers of his special superconducting metals placed onto a plastic board. Some of the problems he had encountered with the speed of the cryotron had been attributed to the comparatively thick wires he was using. By swapping the wires for thin films of the relevant metals, he may be able to improve the speed at which it flipped from zero to one.

The new design was a type of printed circuit board, similar to what you would find inside almost any everyday electronic gadget today were you to crack it open. The process that Buck and his lab partner Shoulders devised to lay down these thin lines of metal carved open the whole field of microchip manufacture.

IN 1950S AMERICA, *Life* magazine was one of America's most widely read publications, selling as many as 13.4 million copies a week. Its high-quality pictures helped to create the genre of photojournalism. Ernest Hemingway first published *The Old Man and the Sea* in *Life*, and the publication also serialized the memoirs of both Harry S. Truman and Winston Churchill.

Everyone read *Life*—including Nikita Khrushchev, the Soviet premier. Fiercely patriotic, it had made its name with its coverage of World War II, publishing photos from the front line of the Normandy landings, for example. It also ran painting competitions for soldiers, awarding prizes of up to a thousand dollars for those it chose to publish. *Life* was such an iconic publication that the Nazis used mock versions of its covers as propaganda during the war. In the Cold War world of the 1950s, *Life* continued to fly the Stars and Stripes as best it could.

In March 1956 *Life* had published an article about technology developments that the Russians were making in an effort to stimulate American interest and investment in technology. In response, James Killian, the president of MIT, had agreed to write an article for the magazine advocating changes to the funding of the American education system—and an ambitious plan to produce more physicists from America's schools. Killian was more than just an academic rent-a-quote spouting his ideas; he was a confidant of President Dwight D. Eisenhower who would be named later that year as the first chairman of the President's Foreign Intelligence Advisory Board.

A photographer from *Life* came to visit MIT for a feature about America's need to train more engineers. Buck was passionate about the subject; he had made it the central thrust of his election campaign to join the board of Wilmington school.

Killian's article, "A Bold Strategy to Beat Shortage," appeared in the May 7, 1956, issue of *Life*—the centerpiece of a sixteen-page spread on MIT's revolutionary approach to developing engineers of tomorrow titled "A Quest for Quality Scientists." As Killian wrote,

> The current scarcity of scientists and engineers has become one of the best advertised shortages of our time, thanks to eager recruiting by employee-hungry companies and the intelligence that the Soviets' educational output in these fields is larger than that of the U.S. Industrialists, educators and public officials are flooded with diagnostic studies seeking to explain the shortage.

So far, unhappily, the diagnoses have been fine but the remedies few—and ineffectual.

It would be folly to suggest that we can find a sovereign remedy for the inadequacies of our educational output of scientists and engineers. We can't. Because we believe freedom of choice more fruitful in the long run, not only for human dignity but for excellence, we cannot do as the Soviets do and tell our young people what to study and what careers to choose. If there are deficiencies in the attitudes and performance of high school students, we cannot change them by fiat. We must use incentives instead of directives. We must marshal public opinion to de-emphasize the hot-rodders among our youth and to encourage hot mathematicians.

The advertisements placed around Killian's polemic tell the story of the hot-rod consumerist society he was addressing. RCA had bought a two-page spread to advertise its Victor Big Color televisions, priced from a staggering $695 to a mind-blowing $995 for a set that came with three speakers and its own walnut or maple cabinet. The carpet maker Bigelow was telling *Life*'s readers that "home means more with carpet on the floor" and that one of its Beauvais Broadloom designs could be bought on installments of as little as $2.65 a week.

The Sealy mattress company was offering twenty thousand dollars' worth of its stock to the *Life* reader who could come up with the best name for the "Sealy Girl"—a coquettish, naked young woman rendered in watercolor. For thirty-one runners-up there was a prize of an "all-expense millionaire's dream vacation for two in Jamaica" with flights from Delta Airlines and accommodation in the Tower Isle Hotel.

There were also ads for Johnson's Car Wax, Your Brand skinless franks (clearly targeted at working mothers), and the Lady Sunbeam shaver ("Start enjoying this new safe way to feminine daintiness").

America was already hooked on the postwar economic boom and the wealth it was creating. Encouraging "hot mathematicians" looked like an uphill struggle, but Killian was unperturbed.

A third of America's high-school students, or about 300,000 kids a year, were going on to college. Another third had no interest in further education. Then there was a group that wanted to go to college but couldn't afford to, Killian said.

Between various state programs, bursaries from corporations, and funds raised by colleges and foundations, America was giving out between $50 million and $60 million a year in scholarships, Killian added. If that could be doubled over two to three years, and then quadrupled over five years, that would take the funding to the appropriate level, he reasoned. If the money couldn't be raised from private foundations, then he proposed a five-year federally sponsored scheme as a "pump-primer."

The cash would be used to create nine thousand new scholarships a year that could be awarded competitively. The first three thousand should be for those who demonstrated excellence in science and engineering, the next three thousand for students that exhibited excellence in any other field of study, and the final three thousand for students who had excelled to such an extent in high school that they could be fast-tracked through their freshman year of college on some kind of advanced program.

Killian went on to caution that the problem might partly resolve itself. There was at the time a "feverish, indiscriminate scramble" to hire good engineers away from universities into commerce, but that could slow. As Killian noted, "There is a danger that some young graduates in this boom market will acquire an inflated view of their importance and will conclude that the normal rules of competition, compensation and accomplishment have been permanently suspended on their behalf. I am sure they haven't."

The surrounding pages were filled with pictures of young MIT students, including some of graduates in job interviews with these same titans of American industry that Killian appeared to be frowning upon.

There were images of MIT students sailing dinghies down the Charles River, mixed with images of stern-looking young men counting on their fingers in an exam room, scrawling calculus on dorm room walls, or experimenting with helicopter rotor blades in a wind tunnel.

On the page before Killian's tirade, there was one particularly striking image, the only image in the whole feature granted a page to itself. It showed a young man holding a vacuum tube in one hand, and a speck of twisted wire in the other that looked more like an insect than a computer component. The caption said it all:

> The look of the future is displayed by Dudley A. Buck, 29, a graduate student working on a research project for his Ph.D. In one hand (left) is a cryotron, the tiny electronic device he is developing. In the other hand (right) is the bulky glass tube it may replace. Rising into view is the vapor of liquid helium, a refrigerating agent in which the cryotron must operate. Since it may lay the groundwork for an electronic computer which occupies only a cubic foot instead of one now occupying 8,000 cubic feet, the cryotron may find a use as a navigational "brain" in the intercontinental ballistic missile.
>
> Suddenly, Dudley Buck was a marked man.

13

THE MISSILE GAP

NIKITA KHRUSHCHEV WAS JEALOUS OF AMERICA'S MISSILES. A voracious reader, the Soviet leader made sure to procure the latest technical papers his KGB agents offered about the status of the US missile program. He saw evidence that the Americans were doing things that his physicists had told him were impossible.

Since rising to power in 1953 after the death of Joseph Stalin, Khrushchev had worked hard to remove the worst extremities of his predecessor—not just by curbing mass killings and gulag deportations but by encouraging an intellectual liberalization of sorts.

He was obsessed with missile technology. So much so that, starting in 1955, he scrapped much of the Soviet Navy's fleet, believing it to be obsolete and pointless. The military hated him, seeing their leader as destructive and shortsighted.

"We cannot say my father was focused on missile technology because it was his passion—it was one of our necessities," explains Sergei Khrushchev, an academic who edited his father's memoirs and now has American citizenship. "It was important for my father and for the Soviet Union, and for all of us, because it was one way to prevent an American first strike. The missile technology gave this possibility to the Soviet Union to get a lead. Especially ICBM missile technology, because it was the only way the United States was accessible to us."

Sergei worked on missile guidance systems himself for many years. "We were more or less on the same level as the United States on missiles, generally. On guidance systems, the United States were some steps ahead. In the engines, we were a little ahead."

Multiple accounts published over the years describe Nikita Khrushchev telling his subordinates that the next war, if and when it came, would be a missile war. A scene in William Taubman's *Khrushchev, The Man and His Era* depicts the Soviet leader lecturing his senior officers in exactly this point. At a missile launch at the USSR's Kapustin Yar weapons development facility in the desert of Astrakhan, a hundred miles or so east of Leningrad, Khrushchev spoke about discarding almost all conventional armed forces to drive as much resource as possible to missile development, according to Taubman. The assembled ministers and generals listened in stunned silence.

In the summer of 1958, Khrushchev was summering in Crimea at the Nizhnyaya Oreanda health resort, where he gathered top officials—including his top rocket scientist, Sergei Korolyov—for informal brainstorming. Korolyov was symbolic of the times. A Ukrainian, he had spent two years in an Eastern Siberian gulag after falling afoul of Stalin, yet now he was in the most exclusive resort in his homeland sitting at the top table with the new leader. He was still not on great terms with the leadership, however.

Khrushchev believed that when it came to missiles he knew more than the top military commanders who would be using the new technology if and when war broke out, according to Taubman. One of Khrushchev's concerns, for example, was that the USSR planned to have its fleet of nuclear-armed intercontinental ballistic missiles (ICBMs) stationed above ground. Surely they would be vulnerable to preemptive strikes by the Americans, the Soviet leader argued. Why could they not be launched from underground silos, hidden from view?

Korolyov insisted this was a nonstarter. The missiles would burn up in the silo due to the heat from the gasses emitted. They would never get airborne. Khrushchev was unconvinced, arguing that the gas would dissipate in the space between the missile and the steel cylinder. Khrushchev realized he had no right to force the idea down their throats, so he let the matter drop. Not long afterward, sometime late in 1958, Khrushchev found a technical report in an American science journal explaining that President Dwight D. Eisenhower's researchers already knew how to launch giant ICBMs from underground silos. After reading this, he supposedly "rejoiced like a child," commanding his missile experts to read every possible scrap of information about America's missile program—and, presumably, the individuals engaged in it.

The state of the Soviet missile program was the central debate in global geopolitics of the day. After World War II, Wernher von Braun, the developer of the V-1 and V-2 flying bombs used to such devastating effect by the Nazis, was lured to America. The Russians, meanwhile, captured a group of Braun's underlings and created the Kapustin Yar facility at which they could develop their ideas. The German scientists were held as prisoners for six years from 1946 until 1951, during which time they trained a group of local Russians to take over their work.

Braun had been cleansed of his Nazi past by American officialdom, in order to facilitate the work they wanted him to do. The public would not have tolerated an ideologically committed Nazi serving in public office. The military was more pragmatic, however, and more concerned with securing the best talent.

Braun and his team were shipped around to different bases, including the US Air National Guard base in New Castle, Delaware; Fort Strong, in Boston Harbor; and the Aberdeen Proving Ground in Maryland. Eventually, by the start of the Korean War, they too had their own dedicated site in Huntsville, Alabama. Braun's team complained, only half-jokingly, that they were "prisoners of peace."

America was not what they expected. Braun had been in charge of thousands of scientists at Peenemunde, on Germany's Baltic coast. Now he was made to answer to Major Jim Hamill, a pimply twenty-six-year-old—who encouraged Braun to avoid even speaking to everyday Americans for fear that his obvious German accent would cause offense. His engineers complained of the "tasteless" American food and begged for linoleum to cover the bare floorboards in their labs.

Whereas wartime Nazi Germany had thrown infinite amounts of cash at his ideas, Braun claimed he was being starved of resources in the United States. He had been developing rockets based on old V-2s that had been seized in Germany and shipped to America. Although he had proposed countless ambitious new rocket designs from the moment he set foot on US soil, and started to talk about conquering space, in the first few years most of Braun's projects were knocked back on the grounds of cost.

It all proved fertile ground for conspiracy theories. Aided by the public musings of the ambitious Democratic Party senator from Missouri named Stuart Symington, the notion of the "missile gap" entered the public discourse. The Russians, the likes of Symington stated, were years ahead of America in development of next generation nuclear weapons that could be launched from the Soviet Union and take out European cities.

With the benefit of hindsight, the consensus view seems to be that Symington—a former chief executive of Emerson Electric—was a paid lobbyist for the American defense industry. Braun, in his attempts to secure more funding for his ever more grandiose schemes, was probably happy to go along with the narrative of paranoia.

President Eisenhower, the former leader of the Allied forces in Europe, knew a thing or two about the relative strengths of his own military and that of the Soviets. He never believed in the "missile gap," nor the "bomber gap" theory that preceded it.

Eisenhower's take on future warfare was identical to Khrushchev's. His New Look defense policies were about spending less money in total, while splashing more money on the most advanced technologies—the science that would win future wars. Yet he was fighting against vested interests left, right, and center, not least from congressmen fearing job cuts at shipyards or munitions plants in their local districts.

Eisenhower knew he had the technological edge. Khrushchev knew that too. In public, the Soviet leader was only too happy to make public statements professing the USSR's military lead. He developed a habit of test-launching missiles at Kapustin Yar immediately before every foreign trip, helping to build his international reputation as a badly-dressed vulgarian with a short temper. In private, he berated his senior officials for falling behind the Americans.

Eisenhower was being subjected to political potshots based on made-up intelligence. His time spent on the golf course came to be associated with complacency about the Soviet threat he refused to acknowledge.

The same narrative would eventually help drive John F. Kennedy to the White House, and then set the scene for the Cuban Missile Crisis. The debate over whose missiles were better, and their respective range, was central to Khrushchev's desire to place missiles in the Caribbean.

Eisenhower's response to the missile conspiracy theories was militarily pragmatic: he would procure evidence to disprove the theories.

In August 1953 the American president asked a favor from Winston Churchill, the reelected British prime minister, requesting that the Royal Air Force fly over the Kapustin Yar research base to see what it could see. "They came back with their planes shot full of holes and allegedly told the Americans that if they wanted that sort of thing done they could jolly well do it themselves," claimed a declassified NSA report by Cecil Phillips and Lou Benson on UK and US efforts to break Soviet intelligence.

The CIA experimented with floating high-altitude balloons, equipped with cameras, across Soviet airspace. Of the five hundred balloons that were sent, only forty-four were ever recovered once they made it to the Far East. Their pictures yielded little information.

Eisenhower was convinced that technology could help him get to the bottom of the problem. In July 1954 he asked James Killian, the president of MIT, to set up a special committee to investigate the available options for surveillance of Soviet missile sites a little more thoroughly. It was dubbed the Surprise Attack Panel, its objective being to prepare against a surprise attack rather than orchestrate one.

Killian invited fellow Bostonian Edwin Land, the inventor of the Polaroid camera—and Virginia Buck's old boss—to join him on the panel.

The solution they agreed on was a Lockheed-built high-altitude plane, based on the fuselage of the F-104 Starfighter, with a specially built Polaroid camera inside. The U-2 spy plane would be dressed up as a civilian project and run by the CIA. On December 9, 1954, Eisenhower signed off on a $54 million order for twenty of the aircraft.

Yet the U-2 could only provide intelligence. It did nothing to advance America's efforts to build its own missiles, or keep them flying straight. When it came to that problem, Dudley Buck was increasingly in demand.

BUCK'S CRYOTRON, AS it was envisaged in its final state, had a particularly clear use for the military. If Buck could build a cutting-edge computer that could fit in a box measuring one cubic foot, he could build a computer that could fit inside a nuclear missile. If he could fit a computer inside a missile, it must surely be possible to use that computer to correct and adjust the missile's path in order to guide a nuclear payload directly to its target.

A properly guided nuclear missile would have been a big step forward. In spite of the propaganda, both the Soviet Union and the United States were still struggling to build missiles that could be directed at a target with any degree of certainty. If either side were to attempt to launch its missiles in anger, who knows where they would have ended up or whom they would have killed.

Buck's cryotron was starting to be talked about among the senior levels of the scientific community at the exact moment that missile paranoia was coursing through the veins of Washington, DC. Just as the U-2 order was being placed, Buck came to Washington to give an updated brief to the NSA about the progress on the assorted cryotron projects.

Before the cryotron was even fully operational in lab experiments, Buck started to be bombarded with questions about its potential use as a missile guidance system.

Experts at the Ballistic Research Laboratories at the Aberdeen Proving Ground in Maryland—one of the sites where Braun and his team were stationed when they first came to the United States—started to drop in regularly to see Buck's lab, diary entries show.

Many of the engineers who had tried to hire Buck over the years had by this time been recruited into the missile effort, including Sam Batdorf, who had been behind Westinghouse's big-money offer to Buck.

As missile technology became the new priority and the area where scientists could secure substantial salaries, Batdorf secured a post as head of electronics at Lockheed Aircraft Corporation's missile systems division in Sunnyvale, California. While his colleagues were working on the U-2 nearby in Burbank, Batdorf was among those at Lockheed trying to build missiles.

As Batdorf wrote to Buck in December 1956,

> I wonder if you could bring me up to date on the status of your work on the Cryotron. As I recall the last time we discussed this you had not yet whipped the problem of its slow speed but had some promising ideas based on evaporation techniques.
>
> The reason I ask this is that we are interested in computers for use in airborne and ballistic missiles. In such applications weight, size and low energy consumption are of prime considerations. On the other hand, of course, all of these factors have to be rated in terms of the amount of computation that can be done per the unit of time by the device in question. We have also the problem of processing large volumes of data on the ground as discussed in the Beacon Hill Studies.
>
> If the present state or the promise of the Cryotron warrants it, I believe we should look together into the matter of exploiting this approach. Therefore, if this turns out to be the case, I would like to be advised as to how we should proceed to get you out here for a conference.

Buck replied that his work was a long way from complete, but was moving along swiftly:

> There is one commercial exploitation of the device being made at Arthur D. Little Inc. of Cambridge, Massachusetts. They have a contract with the Department of Defense to build a large unit containing approximately 250,000 cryotrons. The device is a type of function table which can give a yes or no answer as to whether or not a given 25-bit word is in the unit. The frequency of interrogation will be 100 kilocycles. The unit will occupy one cubic foot exclusive of the liquid helium container and about 2 racks of electronic equipment needed to drive and sense the unit. A feasibility model is at present under construction and indica-

tions look good. The 25,000 flip flops contained in the devices are still of the slow variety. The entire unit requires about 1 and 1/2 minutes to fill from punched paper tape. If, in your missiles systems division you have any use for a cryotron recognition unit of this type, I would suggest you get in touch with Dr. Howard McMahon of Arthur D. Little.

I am confident that it will not be very long before high speed cryotrons will be available. Two companies outside of MIT have already observed superconductors to change from one state to another in less than 0.1 microseconds. One company, in fact, believes they are able to switch superconductors from superconducting to normal in 8 millomicroseconds. I would very much like to keep in touch with you, and if I am out your way this summer, I would very much like to drop in for a visit.

Batdorf's letter inquiring about the cryotron's suitability as a missile guidance system came three months after *Life* magazine had proclaimed to the world that the gadget would eventually perform this role. Buck already knew Batdorf's new boss.

Louis Ridenour was one of MIT's most celebrated alumni, and he had been immersed in classified government projects for many years. He was a big man with a big reputation—known for his ability to crash through the hierarchy of the American military to get projects off the ground. During World War II, Ridenour had led the team in MIT's radiation lab, which developed the SCR-584 radar—a lighter, more accurate radar that helped to sway the course of the war. When it went into service in 1943, it superseded every other radar system in the Allied forces and was carried to the front in specially built trailers that doubled as remote operating stations.

It was Ridenour's radar that had helped Britain fend off the V-1 bombs that were used to attack London in the middle part of the war. Once the radars were installed on the gun batteries of England's south coast, the success rate in shooting down the "flying bombs" rose from about 17 percent to 60 percent. Ridenour was sent to Britain himself to oversee the deployment of his device. He then had a hand in the development of a new radar system that could be installed in a plane, which facilitated bombing raids in dense cloud cover. The system, nicknamed Micky, was installed on the two B-29 bombers that dropped the Little Boy and Fat Man atomic bombs on Hiroshima and Nagasaki.

At the end of the war, Ridenour left MIT to set up the computing department at the University of Illinois. He was also appointed chief scientist to the newly created US Air Force, and the chairman of the NSA's scientific advisory panel on computers. Ridenour wrote the first manual on how to build and operate radar sets—the same book a young Dudley Buck had used as a student when he built his own radar with parts stolen from the Puget Sound Naval Shipyard in Bremerton, Washington, near Seattle.

It was also Ridenour who championed the idea of using private companies to develop America's scientific prowess. He argued that competition, and the profit motive, were powerful drivers to problem solving. Even if there was some duplication of effort, multiple competing teams were much more likely to come up with new ideas than were big, centralized bureaucratic projects like those run by the Russians.

Buck appears to have first met Ridenour in Washington, DC, while he was still working on codebreaking machines. Curiously, Ridenour was granted a leave of absence from the University of Illinois for military reasons at exactly the same time Buck was seconded to work on his covert missions for the CIA. There is a possibility that they were sent behind the iron curtain together on a special project, possibly to negotiate jointly with Konrad Zuse.

What is clear, however, is that by the time Buck had developed the cryotron, Ridenour's name was appearing regularly in his diary. They had a meeting in Atlantic City, New Jersey, a few weeks before Buck got married—a meeting that looks curiously like it may have been his bachelor party. Buck then spent a day with Ridenour at his lab in California while he and Jackie were on their honeymoon.

At the time Ridenour had moved into the private sector. He was working with International Telemeter in Santa Monica—the first company in the world to launch pay-TV services. International Telemeter had an encryption system that could scramble and descramble television signals. The small number of households who signed up for the system bought a box to sit on top of their television set, with a coin-operated meter. For $1.25 they could watch the newest movies or big sporting events.

At its peak in 1954, there were 148 households in and around Palm Springs, California, signed up to the service. Eventually it was forced out of business by the movie studios, who were worried that it would take away theater audiences.

Ridenour coauthored the patent on the company's video scrambling technology. Obviously, a system to encrypt and decrypt signals could have other uses, such as scrambling the pictures that would be sent back to ground from U-2 spy plane missions.

International Telemeter was a hotbed of creativity. Ridenour and three of his colleagues created an early version of the compact disc—an optical storage disc that could store thirty million bits of data by spinning at twenty-four hundred revolutions per minute.

The company was also working on government projects, however. Alongside the assorted TV technologies it was developing, International Telemeter was also contracted to build a computer memory for Wright-Patterson Air Force Base in Riverside, Ohio.

Ridenour was involved in America's missile program from the start. In addition to being chief scientist to the air force, he was also the scientific adviser to the Ballistic Research Laboratories in Aberdeen, Maryland.

If President Eisenhower thought the missile program was off track, it was on Ridenour's head—at least in part.

Shortly after the U-2 spy plane was commissioned, Ridenour took a more hands-on role in the Cold War arms race. He was hired by Lockheed to join its newly created Missile Systems Division. Within weeks of Ridenour's arrival, a large proportion of Lockheed's scientific staff quit to form their own business. Aeroneutronics, the business they created, would go on to work on large contracts related to the US space program.

After this mass walk-out—referred to in Lockheed as the "Revolt of the Doctors"—Ridenour was appointed chief scientist at Lockheed Missile Systems. One of his first jobs was to oversee the transfer of the lab from an existing facility in Van Nuys, California, to a new

THE CRYOTRON FILES | 151

state-of-the-art operation in Palo Alto. It was one of the first commercial research facilities in the area, just a stone's throw away from Stanford University. Lockheed's lab helped to create the foundations of Silicon Valley; the research labs of the modern-day Lockheed Martin are still there today, across the road from the headquarters of Hewlett-Packard.

Once the lab was set up, there was no excuse for not building the perfect missile. Ridenour firmly believed that the reason to build the best missile was to ensure that you would never have to use it. Mutually assured destruction was a solid theory, he believed. He liked to game the Russians, teasing them with bits of technology and throwing a few misleading ideas into the public domain just to keep them on their toes.

"A genuinely secret weapon is absolutely useless in peacetime," Ridenour told *Popular Science* magazine in May 1957. "Not all secrets need to be exposed, and they can be mixed in with a few lies. When a potential enemy can verify some 'secret weapons,' but not all, they will have to be very cautious."

Ridenour had been dropping hints about his idea of the ultimate bomb since the tail end of World War II. Shortly after Hiroshima and Nagasaki had been destroyed (with bombs guided by Ridenour's radar device), he wrote a darkly satirical one-act play, *Pilot Lights of the Apocalypse*, about the accidental destruction of the world. It was published in Fortune magazine—and then syndicated to newspapers across America.

The play is set one hundred feet underground in a "western defense command" center somewhere near San Francisco. It tells the story of two sergeants who inadvertently cause global nuclear apocalypse by mistaking an earthquake for an enemy attack—which their handbook suggests must have originated from a Danish satellite bomb. The reasoning was that Denmark had the highest negativity rating, due to local protests over a statue that had been gifted from the US President to the Danish King.

It is astonishing to think that such a biting polemic against the dangers of the nuclear bomb could have come from the pen of someone with a hand in its design and possible deployment. That Ridenour was kept in the military fold in spite of this public attack is even more extraordinary.

Yet the satellite bombs that Ridenour predicted in his satire were more than just science fiction. By the time Ridenour arrived at Lockheed, Wernher von Braun had been working on such a theory for years. Launching missiles from a satellite attack platform was one of the ultimate goals of what he was trying to achieve. As far back as 1946 Braun had warned the US Army that the "nation that first reaches this goal possesses an overwhelming military superiority over other nations."

In an article for *Collier's* magazine in 1952 Braun had stressed again the peril in allowing the communists to dominate space with manned space platforms full to the brim with nuclear weapons. He also outlined how the United States could build a circular space station orbiting Earth that could be used as a launchpad for manned missions into deep space. The editor's comment in *Collier's* read, "What Are We Waiting For?"

Walt Disney subsequently made a TV show about space exploration, extrapolated from Braun's article.

With the Cold War at its peak, no ideas were off the table. Ridenour wanted Buck's help to get some of those ideas off the ground— quite literally.

14

FAME

Fʀᴏᴍ ᴛʜᴇ ᴍᴏᴍᴇɴᴛ Dᴜᴅʟᴇʏ Bᴜᴄᴋ'ꜱ ᴘɪᴄᴛᴜʀᴇ ᴀᴘᴘᴇᴀʀᴇᴅ ɪɴ *Lɪꜰᴇ* magazine, reporters from all over America started to turn up unannounced at MIT, on the hunt for the famous professor behind the "tiny electronic brain." Even though Buck was still a long way from satisfied with its performance, the hype about the cryotron had become self-sustaining.

Initially Buck seemed to enjoy the attention—as did his cash-strapped researchers. As Chuck Crawford remembers,

> Particularly after that *Life* article, Dudley was world famous in this area. These reporters would come by—I would be sitting at my bench. Somehow they would get into the institute, find the right place, and come in to knock on Dudley's door. The door was always open, he was open for everybody. So they would go in and talk for a little bit.
>
> The reporter would then invite Dudley to lunch, which was great for him because he could get quotes he could use about the future of computers that could turn into an article for him. Dudley would say, "Is it okay if I bring a couple of staff members with me?"
>
> The reporter is trapped at that point, as you can imagine. He says, "Yes, of course, Doctor Buck." Dudley would then get out of his chair, walk over to his office door, step just outside, and then he would hold his right hand up and put two fingers in his mouth and make this tremendous whistle. He could do a very loud whistle.
>
> Everybody would then pile out—they knew this meant free lunch. This poor journalist has now got six or eight people to take

to lunch. Dudley would give him a few quotes—he would give him something for his money.

By now, Buck had a sizable crew of technicians. They rigged up some experiments that served as good party tricks to demonstrate the invention. According to Don Burrer, one of the students working there, Buck rigged up a set of cryotron switches to some sound components so that it could play simple musical tunes for the admirers who came to the lab.

The media attention was followed by serious scientific recognition. In January 1957 Buck received a telegram from the Institute of Radio Engineers to tell him he had been awarded the Browder J. Thompson award—a special medal recognizing achievements by scientists under thirty. It was named after a former General Electric scientist who had invented vacuum tubes and was killed in World War II while operating a radar on a plane that got shot down over Italy on a night flight. The prize had been awarded for the paper Buck had presented to the institute the previous spring, explaining the full working detail of the cryotron. He was to be honored at a lavish banquet in New York City that March, the telegram told him.

The trickle of interest then turned into a torrent. Suddenly the cryotron was on the front page of *The New York Times*, illustrated with a close-up shot of Buck's hand, holding a vacuum tube, a transistor, and a tiny Z-shaped piece of wire that looked like a splinter.

"Devices so small that 100 of them can fit in a thimble may open the way to a revolution in electronic computer design, according to the engineer who developed them at the Massachusetts Institute of Technology," the newspaper said. "The devices, called cryotrons, suggest the possibility that computers occupying only a cubic foot of space may do the work of machines that now fill whole rooms."

The electronics trade press was even more excited, pumped up by the Arthur D. Little publicity machine: "$1,000 Computer Coming," screamed a headline in *Electronics* magazine; the article claimed, "In 10 years, research in cryogenics may make possible $1,000 versions of computers now costing $1 million. . . . Key to a $1,000 computer may be an information storage device based on the science of cryogenics, called the cryotron." The article went on to explain how the switching speed of a cryotron was expected to increase a hundredfold by further research and development, and that it may also improve radar reception by 100 percent.

None of the reports mentioned the cryotron's possible role as "navigational brain" for nuclear missiles, however, as it had been previously described in the *Life* article. Although this potential military application was presumably a key reason behind the interest in Buck and his device, that information was missing from the coverage.

Radio stations and TV networks wanted to talk to Buck. After a prime-time appearance on one of the Boston TV stations, he became a local celebrity, and the Buck family's neighbors finally understood what Dudley did for a living.

Everybody wanted a piece of him. Howard Rodman, the screenwriter who would go on to create *The Six Million Dollar Man*, came to MIT to meet with Buck about the possibility of dramatizing his story for CBS. It seems that nothing ever came of it, but the meeting was logged in Buck's diary.

The US Information Agency Broadcasting Service sent Buck two half-hour rolls of film and a long list of questions, coupled with instructions on how to record and interview himself. It's not clear from the correspondence who the film was intended for or how widely it was to be distributed.

MIT seems to have been caught off guard by the level of attention Buck had started to receive. The university gave him airtime on the campus radio station, but all the exposure had highlighted an embarrassing failing on the part of the department of electrical engineering.

Although he was being called Professor Buck informally by his students, MIT had still not yet awarded him his doctorate. It appears that he was led to believe by his thesis adviser that his work on the cryotron would not be sufficient to qualify.

Gordon Brown, the Australian-born head of the electrical engineering department, took the matter up with Jerome Wiesner, the dean of science, who would go on to become MIT president. Brown and Wiesner had worked together at MIT throughout the war. The problem, Brown claimed, was that Buck was being held back by

Arthur Von Hippel, another highly decorated MIT veteran who had made his name developing radar technology.

Brown wrote,

> Dudley Buck's status with respect to thesis is a matter that bothers me greatly. At the present time Dudley is under the impression that he still has ahead of him his Doctorate thesis research. He is somehow reporting to Dr. Von Hippel on this matter. I consider the whole affair ridiculous and worthy of immediate action by the Graduate Committee.
>
> The facts are these. Dudley has been recognized for some time as a rare and extremely capable experimentalist. He has achieved considerable recognition in the professional world for a fine piece of work. He has been recognized by the top professional society in the communications field. He has been the victim of a great deal of publicity in the interests of publicizing MIT. It looks to me as though, unless something is done, he will be two or three years completing for the Doctorate. Even this will be one of the shortest Doctorate theses that anyone has ever achieved under Von Hippel's guidance.
>
> I think a special committee should be formed with the clear authority to extract from Dudley a manuscript that reports on what he has done, get it typed and have it submitted before June 1957 as a doctorate thesis.

MIT decided to lighten Buck's academic teaching load to help fast-track his doctorate, with several professors acknowledging the "embarrassing" situation the university had gotten itself into.

Now that there was a degree of pressure from MIT to prioritize his academic paperwork, the media attention was starting to become a little tiresome. The timing of his newfound fame was also a little inconvenient. In December 1956, a month before he learned of his award from the IRE, Jackie gave birth to their second child—a baby boy named Douglas who weighed in at a whopping eleven pounds and one-half ounce after arriving almost five weeks overdue. Carolyn, Buck's daughter, was only fifteen months old when her brother was born.

Whereas Buck had awaited the birth of his first child by devotedly stalking the hospital corridors, Douglas arrived with less fanfare: based on diary and lab book entries Buck spent most of the day at work, advancing his cryotron experiments and talking on the phone about problems with assorted NSA projects.

Now that he had become an internationally recognized figure, there were additional constraints on his time. After the press reports and the TV appearances, letters started to flood into Buck's office from all over the world, soliciting business, offering work, or seeking information.

W. Nijenhuis from the Dutch electronics giant Philips flew from Eindhoven to meet the creator of the famous cryotron. Professors from France, Italy, and Uruguay wrote to demand papers and information. Some of the British newspapers then caught on to the story. Old friends from the V-12 program got in touch. One university friend, who had become a teacher in Menlo Park, California, joked that he would be "looking for a computer that fits in the vest pocket, with an A.D. Little liquid air machine trailing behind."

The most bizarre request came from a leather tanning firm in Barcelona. While the letter was cryptically addressed to "Dudley Buck, engineer, Institute of Technology, Massachusetts," it somehow reached him:

Dear Dudley,

We have heard that you have succeeded in the production of a small electronic brain.

As our firm is very interested to be informed as wide as possible with regard the possibilities to introduce the said machine in our market, we should be very glad if you would be so good to get us in touch with manufacturers of the same.

Thanking in advance for your kind reply, we remain Dear Sir
Yours faithfully
Comercial Serra

Buck sent an exasperated reply, telling the Spaniards that "the newspaper coverage of our research has been a bit misleading" and that there was nothing yet to sell them.

Albert Ducrocq of the Société française d'électronique et cyberné-tique got a more comprehensive dismissal, coupled with a copy of Buck's award-winning paper:

> The large amount of publicity which appeared in recent news-papers was rather embarrassing. Actually we do not have a computer using cryotrons under construction at MIT, but rather we are studying the fundamental properties of super-conductors and cryotron circuits. The Arthur D. Little Com-pany of Cambridge is engaged in building a unit containing a large number of cryotrons for the United States Government, but this is not properly called a computer. In fact, it could not be properly called a computer memory either. It is a very spe-cial device which gives one a YES-NO answer as to whether or not a, let us say, book is in his library.

The example of finding a book in a library combined two Buck inventions: his content-addressable memory system that governed the filing process and the prototype cryotron circuit that was providing the computing power. Yet Buck told Ducrocq that, contrary to the in-formation in MIT's press releases, the cryotron would not be in a po-sition to replace the transistor in computers of the near future. It was all a project for "some years from now."

Buck's foreign correspondents all got the same tale of complica-tions, drawbacks, and delays. When American companies like Boeing, IBM, or RCA got in touch, however, Buck told a rather different story. He still cautioned that the cryotron was less advanced than they all hoped. Yet he would go on to elaborate about the cryotron-based re-search programs that had been launched all over the country—many supported by funding from the NSA. One of Buck's recent graduates who had been hired by General Electric had already developed a cry-otron that was switching so quickly that his equipment could not measure the speed.

The cryotron was progressing about as well as could be expected. Yet Buck was now reticent to indicate just how quickly it was advanc-ing to any party outside the broad sweep of NSA influence.

COMPUTERS WERE STARTING to drift into the popular consciousness of America. While Buck had felt it necessary to spell the word to his wife, Jackie, and her college friends only five years earlier, the success of the likes of IBM and Remington Rand in selling their machines had popularized the arrival of the digital age.

The vision of the future that Buck had been portraying for years was starting to become a reality. Computers were no longer the preserve of military codebreakers.

The US Census Bureau was the first big computer user, buying a Universal Automatic Computer (UNIVAC) to keep track of the American population. By the time Buck shot to fame, big corporations were starting to use computers to handle their accounts and run their payroll systems.

As the *Wall Street Journal* wrote in a lengthy article on the burgeoning computer industry on August 15, 1957,

> An electronic hum, a clackety-clack of robot typewriters, a few red lights blinking—and Sears, Roebuck workers pick up their pay checks every week. The hum-clack-blink routine is a giant electronic computer grinding out weekly a massive payroll that Sears couldn't complete in less than two weeks back in the old pre-computer days, even using a small army of clerks. It goes without saying that the more frequently paid Sears folk greatly admire the work of the electronic brain they seldom see.
>
> The whole computer world, in fact, is basking in general satisfaction, almost to the point of smugness. Computer users, including retail stores, steel mills, drug manufacturers and insurance companies are saving time and money by feeding an increasing variety of chores to the machines. And computer makers can hardly keep up with demand for their brainy products; the value of computers sold or rented may hit $350 million this year, nearly four times last year's $94 million figure and all but out of sight of the $47 million 1955 total."

The typical computer at the time cost up to $1 million to buy, or as much as $50,000 a month to rent, the article explained. They were different from the previous generation of data-processing machines mostly due to their speed and their ability to handle vast amounts of information—credited largely to the "magnetic core" memories that Buck had helped create.

The machines were far from perfect, the article pointed out. Long Island Lighting Company had sent a monthly electric bill for $1,266.80 to one family. "The householder, one Raymond Carr, didn't think he'd been that careless about leaving the lights burning and complained. Long Island Lighting re-checked, evidently using a slow but trusty pencil, and sent Mr. Carr another bill for $11.83."

Then there were examples cited of other companies that were now able to perform calculations they had never thought possible previously. Great Lakes Pipe Line spent five and a half years and $140,000 developing a computer to handle its network. The number of products being shipped through its pipeline had quadrupled in the previous twenty years, the company complained. Thanks to its new computer, the company could have up to two hundred different types of oil or other liquids sloshing through its various pipes at one time, which could be loaded or unloaded at sixty different points along the way. Before the computer, the job had been done with "a great deal of intuition," the company confessed.

Then there was Commonwealth Edison in Chicago, which had bought a computer to help it buy coal more effectively. By tracking the quality, source, and transportation costs of different types of coal available for sale in the market, the company was supposedly shaving tens of thousands of dollars from its annual energy bill. The system must have been one of the first commodity trading programs in existence.

Abbott Laboratories, meanwhile, was using a computer to track its inventory. The giant pharmaceutical company told the *Wall Street Journal* that while it spent $60,000 a year on computers, it had saved $40,000 by stopping the rental of other tabulating machines and another $20,000 in salary costs. The big payoff, however, came in the fact that it could carry less inventory in each of its warehouses because it knew exactly what was stored where.

America's big insurers, Allstate, MetLife, and State Farm were all quoted in the article, bragging about the cost savings they were seeing from their new computers.

Left: Dudley and Virginia Buck
with their mother.
(Credit: Buck family archives)

Above: The Buck family in Santa
Barbara, California. *From left to right:*
Grandma Delia, Grace, Virginia,
Dudley, and Frank.
(Credit: Buck family archives)

Left: Buck and his Aunt Gladys
photographed on his graduation
from the V-12 program,
July 7, 1947.
(Credit: Buck family archives)

An autographed photo of Scout Troop 31, led by Scoutmaster Buck, with
J. Edgar Hoover.   (Credit: Buck family archives)

HEADQUARTERS
7821 COMPOSITE GROUP
APO 407, US ARMY

A7.4 SDE   25 April 1950

SUBJECT: Concealed Weapon

TO: Whom It May Concern

        1. The following named US Civilian, whose signature appers be-
low, is authorized to carry a conceal weapon while in the performance
of his duty for the peroid beginnin 25 April 1950 to 5 May 1950.

        Mr Duddley Buck GS - 14   Colt .32 Automatic SN.561180
                                       US Property

        2. Authority: Per telephone conversation Provost Marshal, Munich
Military Post, APO 407-A, US Army and Commanding Officer this headquarters.

        FOR THE COMMANDING OFFICER:

OFFICIAL: EDWARD BRAY
                                            Captain, CAV
                                            Adjutant

EDWARD BRAY
Captain, CAV
Adjutant

        Tel: MMP EX 28 or 35

HEADQUARTERS
7821 COMPOSITE GROUP
APO 407, US ARMY

A7.4 SDE   25 April 1950

SUBJECT: Concealed Weapon

TO: Whom It May Concern

        1. The following named US Civilian, whose signature appers be-
low, is authorized to carry a conceal weapon while in the performance
of his duty for the peroid beginnin 25 April 1950 to 5 May 1950.

        Mr Duddley Buck GS - 14   Colt .32 Automatic SN.561180
                                       US Property

        2. Authority: Per telephone conversation Provost Marshal, Munich
Military Post, APO 407-A, US Army and Commanding Officer this headquarters.

        FOR THE COMMANDING OFFICER:

OFFICIAL: EDWARD BRAY
                                            Captain, CAV
                                            Adjutant

EDWARD BRAY
Captain, CAV

Two official letters from the US Army signed by Captain Edward Bray granting
permission for US Civilian Dudley Buck to carry a concealed weapon in Germany
"while in the performance of his duty," on two separate periods in the spring of 1950.

(Credit: Buck family archives)

Dudley and Virginia Buck and Glen Campbell move to Boston, Massachusetts, 1949.

(Credit: Buck family archives)

ELECTRICAL ENGINEERING JUNIOR STAFF RATING FORM

CONFIDENTIAL DATA. At Professor Hazen's request, no Junior Staff member will have access to this information.

TO: Mr. R. A. Plachta
    Rm. 4-203A

NAME: Dudley A. Buck   RANK: Research Assistant   AGE: 24

Received BS degree on   June 1948   from   University of Washington
                          (Date)   (College)

Studying for   M.S.   degree at MIT. Will receive his degree on   June 1952
                                                                    (Date)

I. Rating of man's performance: CHECK ONE

   [ ] Superior   [x] Good   [ ] Satisfactory   [ ] Below average

II. Status Recommendation: CHECK ONE

   [ ] Reappoint for 1 year and promote to _____
                                                      (Rank)
   [x] Reappoint for 1 year.

   [ ] Do not reappoint. Plans to leave E. E. staff in June.

   [ ] Do no reappoint. Performance unsatisfactory. Explain below.

III. Salary Recommendation: CHECK ONE

   [ ] Maximum salary increase   [ ] Minimum salary increase

   [x] Moderate salary increase   [ ] No change in salary

IV. Statement by supervisor evaluating man and his performance on job:

   Buck has worked during the past five months principally on design
and construction of interim display equipment for WWI. He is an eager
worker with a great deal of initiative. He has a very agile mind and
executes his ideas rapidly, sometimes a little too rapidly. Has good
judgment, although it is sometimes postponed because of haste to build.

   He has a pleasant disposition and is a cooperative worker.

From Folder Lincoln Lab.

1/2/51
_____
Date

_____
Signature of Supervisor

A copy of Buck's electrical engineering junior staff rating, dated January 1951—
an overall positive evaluation noting "good" performance and recommending a
moderate salary increase.   (Credit: MIT archives)

The Cryostat, a machine built by S. C. Collins at MIT to liquefy helium. Experiments with the helium produced by this machine revealed that some metals became superconductors for electricity once they were steeped in a vat of liquid helium; this revelation inspired Buck to begin his research into using these superconducting materials to create a miniscule and ultrafast computer, leading to his invention of the Cryotron.   (Credit: Buck family archives)

Buck and Jackie Wray on their wedding day,
June 6, 1954.   (Credit: Buck family archives)

Buck and Jackie attend the wedding of Buck's sister, Virginia,
and Georg "Schorry" Schick, September 4, 1954.

(Credit: Buck family archives)

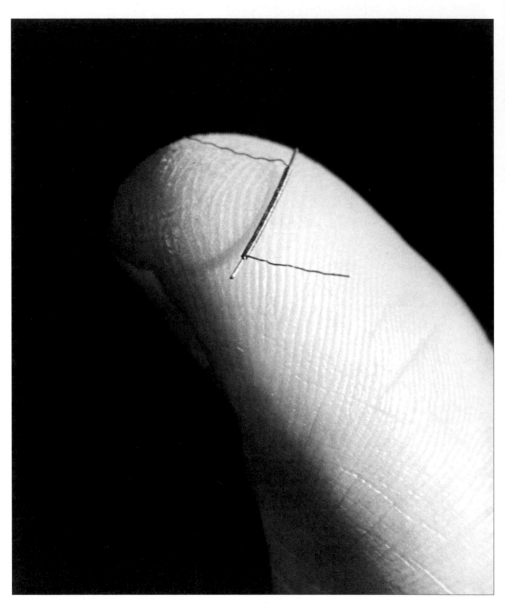

A single cryotron. On February 18, 1954, Buck had his first successful test of the cryotron, in which the device switched from conducting an electrical current to resisting it. (Credit: Buck family archives)

A cryotron, a "superconducting switch,"
could be hooked up to a circuit.
(Credit: Buck family archives)

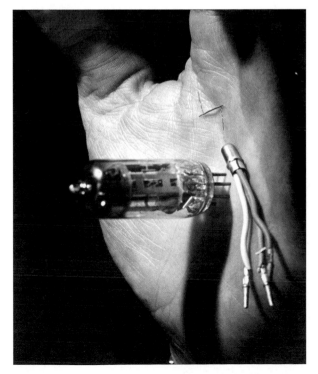

(Credit: Buck family archives)

Magnetic core memories, developed at MIT by Buck and his colleagues. Magnetic core memories allowed data-processing machines to handle vast amounts of information and were used in IBM computers in the 1950s. The invention of this technology led to a legal battle between IBM and MIT.

(Credit: Buck family archives)

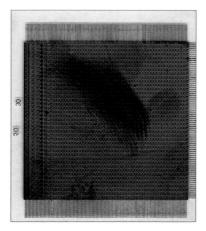

By 1956, Buck and his colleague, fellow MIT engineer Ken Shoulders, had developed his cryotron from two thin wires to a thin film made of super-conducting materials, which increased the speed at which the switch could flip between "zero" and "one."

(Credit: Buck family archives)

HE CHRISTIAN SCIENCE MONITOR

AN INTERNATIONAL DAILY NEWSPAPER

COPYRIGHT 1957 BY
THE CHRISTIAN SCIENCE PUBLISHING SOCIETY

BOSTON, WEDNESDAY, FEBRUARY 6, 1957   * * ATLANTIC EDITION   TWO SECTIONS   FIVE CEN

traits

Museum embrandts

By Dorothy Adlow
Art Editor of The Christian Science Monitor

most ently so well-to-do that he could commission these likenesses by Rembrandt, who had already have gained a reputation. In 1634, the artist was 28 years old. tifferend rend

Soft Neutrals Used

by The Elison portraits were ting, planned as companion pieces. Husband and wife are posed at can-three-quarter view, facing each the other. Books and manuscripts are heaped on a table emblemize easily the calling and the learning of aures the clergyman. His facial ex-lec-pression is appropriately serious sion, and meditative. A rosy glow and illuminates the well-defined fa-ride cial features. Both face and andt hands provide contrast with the dark robes broadly painted by cific the expert, assured hand elite Maria Elison accommodates cant herself with appealing modesty to this unusual assignment, sit- of ting for an eminent artist. Rem-tri. brandt painted her as though he liked and respected her. No ared adjuncts, no accessories enlarge of the context of Maria's portrait. andt She poses against a background as is illuminated with soft neutrals, except for the ornate, stiff ruff which was the fashion of the day. The large hat which frames her features adds a stylish note to what could have been a se-rious matronly characterization.

Gallery Festooned

Other paintings by Rembrandt in the Museum collection are "Lady with a Gold Chain," "Man in a Black Hat," "An Old Man," "The Artist in His Studio," and "St. John, the Evangelist." The first two are a companion pair painted in 1634, the "St. John" is a work of Rembrandt's ma-turing years. In the Isabella Stewart Gard-ner Museum is a double portrait by Rembrandt, "A Lady and Gentleman in Black." This is a work of the same period, 1633, and the married couple wear costumes similar to the garments of the Elisons. The two masterpieces have

Tiny Device Replaces Tubes or Transistors in Giant Computers

Called a "cryotron," this element consists basically of just one wire wrapped around another. Used in electronic computers, it prom-ises to reduce drastically the size of these machines and expand their capacity far beyond what is possible today. It is held by the inventor, Dudley A. Buck, a student at the Massachusetts Institute of Technology. [Story: Page 6.]

No Barter for Prisoners

Dulles Bars Red China Deal

By Neal Sanford
Staff Correspondent of The Christian Science Monitor

Washington

American newsmen—some at least—want to visit Communist China. Should the United States help them to do so?

There are more angles to this seemingly simple issue than to

ter—until either he or the President amends them."

The secretary asserted that, Peking permitting newsmen to enter Communist China in return for the release of 10 Americans still imprisoned

munists make the connection," Mr. Dulles declared, "we cannot escape the consequences of that connection or escape the fact that, if we give in to it, it puts a premium for all time and at all places upon seizing and impris-oning Americans illegally and

Negotiation On Algeria In UN Posed

By Mary Hornaday
Staff Correspondent of
The Christian Science Monitor

United Nations, N.Y.

Eighteen Arab-Asian nations have proposed that the United Nations "invite" France to nego-tiate with the Algerians with UN assistance.

This has been done over the protest of United States Secre-tary of State John Foster Dulles who feels that it is difficult to put the Algeria issue into a UN resolution. The move counters, also, the plea of France for a supernationalistic "Eurafrica" approach.

Mr. Dulles said in a Washing-ton press conference Feb. 5, "I do not know whether it is pos-sible or would be an advantage to try to arrive at any substan-tive resolution (in regard to France and Algeria) which would be voted upon, and per-haps it would be better and perhaps would help the whole situation if that was not at-tempted. These issues are very complicated, and it is not very easy to put them in the frame-work of a resolution."

Mr. Dulles' statement was projected against the back-ground of hard political facts that if the Algerian question went put to a vote, the United States would be stuck on the horns of the dilemma either of slighting the French or t' Arab-Asian bloc.

France Protests

The text of the proposed resolution was put before the UN's Political Committee over the objections of French Foreign Minister Christian Pineau, who insists that UN consideration of the Algerian problem is a vio-lation of Article 2, Paragraph 7 of the UN Charter which for-bids the UN to "intervene in matters which are essentially within the domestic jurisdiction of any state."

More conciliatory to lose than other Arab approaches to the Algerian question, the resolu-tion calls on the UN to "recog-nize the right of the people of Algeria to self-determination

President Fi In Suez St

By William H. Stringer
Chief, Washington News Bureau, The Christian S

President Eisenhower and Secretary of Stat Dulles hope to solve the problems of Suez wit hard on anyone's toes—and that means the Egyptians, the big oil companies, the Texas Texas Railroad Commission.

That won't be easy.

The United States hopes everyone will act wi self-interest and observe international law an would make Suez solutions and oil shipments But it's a lot to hope for, even when Washin its prestige and persuasion on that side.

The President was asked at his news confer the United States would be prepared to supp against Israel if that country refuses to withdraw its forces from the Gaza strip and the Gulf of Aqaba.

U.S. Treads Softly

He carefully replied that Israel was established by the United Nations, that there was a UN resolution asking it to with-draw its forces, and that he per-sonally believed Israel has "a decent respect for the opinions of mankind" and would with-draw.

Mr. Dulles he carefully avoided any direct mention of United States support for sanctions. The United States, he said, is doing all it can to reach a Middle East settlement covering the whole range of complaints, not merely what Israel should do.

A correspondent for the Na-tional Jewish Post reminded Mr. Eisenhower that since 1951 Egypt had disregarded a United Nations determination that all nations, including Israel, should have free passage through the Suez Canal.

But the President and his Sec-retary of State obviously do not want to say anything now that will upset the new-found friend-ship with the Arab world. Mr. Eisenhower has just had ex-tensive talks with King Saud of Saudi Arabia, and told his news conference that he believes the Arabian monarch has a good understanding of the Eisenhower Doctrine—of what the United States proposes to do in the Middle East. Immediately after

steps to fill the by conference a and then, "if other region, of own facilities to get the proper abroad."

This was the a of federal inter the President ha ington still hope try will act by it distance sees the pect that the Sue cleared for shipp of March.

Overproducti

Even while the press conference ent Secretary of lix Wormser up. Mr. Wormser Texas Railroad regulator of Tex tion, undoubtedly state's producer caught with when the oil cris it Texas boosted overmuch now.

Mr. Eisenhow asked at his pr about the sudder prices in the mid of shortage. The plied that Defe Arthur Flemming ported to him on sons" for that ris

Professor Buck and his undergraduate research assistants, including Allan Pacela and Chuck Crawford, in his lab at MIT's Building 10 in the summer of 1958.   (Credit: Supplied by Allan Pacela)

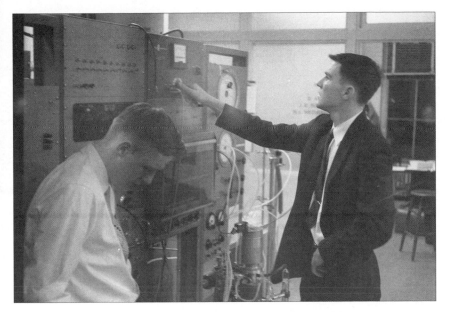

Buck and Chuck Crawford in the Building 10 lab at MIT.
(Credit: Supplied by Allan Pacela)

Buck receives his doctorate,
here pictured with his wife,
Jackie, and children Douglas
and Carolyn, 1958.
(Credit: Buck family archives)

NATIONAL SECURITY AGENCY

FT. GEORGE G. MEADE, MD.

Serial: EXSAB 133-59
29 April 1959

Mr. Dudley A. Buck
10-397A
Massachusetts Institute of Technology
Cambridge, Massachusetts

Dear Mr. Buck:

Dr. Ridenour, Chairman of the NSASAB Panel on Electronics and Data Processing is planning to hold a meeting of the panel on Monday and Tuesday, 25 and 26 May 1959, unless this date would be impossible for too large a proportion of the panel membership.

The meetings will be held at the NSA site at Fort George G. Meade, Maryland.

I would appreciate hearing from you as to whether you will attend. Travel arrangements may be made if you will notify us as soon as possible of your preference. If it is more convenient to personally arrange for travel, please obtain receipts and you will be reimbursed later.

Sincerely,

S. Kullback

S. KULLBACK
Executive Secretary
NSA Scientific Advisory Board

Copy furnished:
Dr. Louis N. Ridenour

A letter (dated 29 April 1959) from Solomon Kullback, Executive Secretary for the NSA Scientific Advisory Board, inviting Buck to attend a meeting on electronics and data processing, hosted by Dr. Louis N. Ridenour at the NSA site.

(Credit: Buck family archives)

MILESTONES

Born. To Margaret Truman Daniel, 35, daughter of ex-President Harry Truman, and Elbert Clifton Daniel Jr., 46, assistant to the managing editor of the New York *Times*: their second son; in Manhattan. Name: William Wallace.

Born. To Mary Churchill Soames, 36, youngest daughter of Sir Winston Churchill, and Christopher Soames, 38, British Secretary of State for War: their fifth child, third son (Sir Winston's tenth grandchild); in Eridge Green, England.

Divorced. Mickey Rooney, 38, pint-sized (5 ft. 3 in.) Hollywoodian admirer of taller girls (his first three wives: Ava Gardner, 5 ft. 5½ in.; Betty Jane Rase, 5 ft. 7½ in.; Martha Vickers, 5 ft. 4 in.); by Elaine Mahnken, 29, 5 ft. 5 in. ex-model; after six years of marriage, no children; in Santa Monica, Calif.

Died. Dudley Allen Buck, 32, exuberant M.I.T. electrical engineer and miniaturization expert, who developed the tiny cryotron to replace the transistor, was working on a cross-film cryotron (diameter: four-millionths of an inch) that would reduce a computer from room to matchbox size; of virus pneumonia; in Winchester, Mass.

Died. Louis N. Ridenour Jr., 47, top-notch nuclear physicist who, despite being emotional about his specialty (in 1946 he wrote a grim, prophetic, one-act play about flocks of satellite bombs orbiting 800 miles above the doomed earth), pioneered in missile programs as chief scientist (1950-51) of the Air Force, helped develop the Polaris and X-17 missiles as research director of Lockheed Aircraft Corp.'s missile-systems division, became a Lockheed vice president last March; of a brain hemorrhage; in Washington.

Died. Oswald D. Heck, 57, popular, powerful, longtime (23 years) Republican speaker of the New York state assembly, who ruled the often unruly legislators with fair play and wit, pushed through controversial measures (State Commission Against Discrimination, compulsory auto insurance, Governor Nelson Rockefeller's tough tax program); of a heart attack; in Schenectady, N.Y.

Died. Carl Holderman, 65, longtime (1918-54) New Jersey union organizer, once described as "the movie idea of a genial Texas oilman"; of a heart attack; in Newark. Holderman was an early C.I.O. organizer, later headed the New Jersey C.I.O., was appointed state commissioner of labor and industry in 1954 by Governor Robert B. Meyner, cleaned house at the scandal-ridden labor department.

Died. John Foster Dulles, 71; of cancer; in Washington (*see* NATIONAL AFFAIRS).

Died. Apsley Cherry-Garrard, 73, polar explorer who accompanied Robert Falcon Scott on his fatal Antarctic expedition in 1911, later described in chilling detail (*The Worst Journey in the World*) a side trip he and two companions made to find emperor penguin eggs; in London.

Died. Irakly Tsereteli, 77, leading Social Democrat who returned from Siberian exile at the outbreak of the 1917 Russian Revolution, served as Minister of the Interior in Kerensky's provisional government until Lenin and the Bolsheviks ran him and all other moderates out of power; of cancer; in Manhattan.

Died. Stephen L (for nothing) Richards, 79, first counselor in the First Presidency of the Church of Jesus Christ of Latter-day Saints, one of the powerful triumvirate that rules the church, shrewd lawyer and businessman who concentrated on the far-flung Mormon missionary program; of a heart attack; in Salt Lake City.

Died. Sir David William Bone, 84, British master mariner who went to sea at 15, commanded troopships under fire in two wars (last to leave the torpedoed transport *Cameronia* in World War I, he grabbed the stay of a destroyer alongside as his ship sank), wrote several books about the sea (*The Brassbounder, The Queerfella*); in Farnham, England.

TIME reports on the deaths of Dudley Buck and Louis N. Ridenour, June 1, 1959.　　　　　　　　　　　　　　(Credit: *TIME* magazine)

Dudley Allen Buck, April 25, 1927–May 21, 1959.   (Credit: Buck family archives)

IBM confessed in the piece that demand for computers had been "considerably greater" than even it had expected when it had started its sales effort in earnest five years earlier. In spite of the huge range of applications described in the article, there were still only about two hundred large-scale computer systems in operation in America, the *Journal* said.

A. N. Seares, the vice president for management services at Remington Rand, said that "the estimated market potential in the U.S. and Canada for large-scale computers is more than 1,200 systems by 1960 and 1,500 by 1965." He was only referring to industrial demand, ignoring the incalculable needs of government and the military.

A 600 percent increase in the size of the market over just five years was a bold prediction. One of the key restrictions to the market, other than the cost, was the amount of space needed to house a computer. A typical commercial computer at the time, such as the IBM 700 series machines, still needed about two thousand square feet of office space.

That was partly why Buck's cryotron generated so much excitement. It was very easy to comprehend why a small computer would be of benefit: computers at the time were inordinately large machines. In media terms, one could also assume there was an appetite to find new heroes from inside this technological world of the future to help sate the public interest in computer technology, which was being fueled by science fiction writers like Isaac Asimov as well as the dawn of the space race.

Buck had a pithy wit and easygoing manner. Yet he also had a much rarer talent for a scientist: he could explain his work in layperson's terms. His easily quotable predictions for what the cryotron would do ensured that more and more people wanted to talk to him. Given the novelty of computing technology and the general level of ignorance about it, Buck's ability to break down this complicated new area of research would have been equally useful to senior military officers.

BUCK'S CONSULTANCY CONTRACT with the NSA had been renewed in the summer of 1956, with a glowing recommendation from Solomon Kullback and a new fee of forty-five dollars a day for every day worked.

Speaking engagements around the country gave Buck an alibi for his countless trips to conduct assessments of NSA projects around the country. Some took more time than others.

Project Vanguard, the code name for America's attempts to launch a satellite, became one of the assignments that required Buck to have a material involvement—albeit somewhat laterally.

In July 1955 President Dwight D. Eisenhower announced that he planned to launch a man-made satellite into orbit around Earth as America's contribution to the International Geophysical Year. The American public was largely disinterested, but large parts of the science community sparked into life.

Satellites had been considered a slightly offbeat area of science until this point. Senior figures in the American defense establishment were opposed to the idea of devoting so many resources to something that sounded fantastical. At this early stage, the objective was solely to put something into orbit. The purpose in doing so was unequivocally military.

After Radio Moscow announced in January of 1955 that the USSR would be launching a satellite soon, the issue drew more attention in Washington. Most of the science community was skeptical that the Russians were telling the truth, but Eisenhower's advisers decided to kick-start the American program nonetheless.

Whatever happened, the United States had to ensure that its Cold War enemy could not claim ownership of space, as Wernher von Braun had warned.

Braun and the other scientists working on long-range missiles had done extensive research on the technicalities. They had already calculated, for example, the velocity a projectile would have to reach in order to escape Earth's atmosphere at various angles and trajectories.

A handful of satellite feasibility studies had been conducted since the tail end of World War II, but there were still a great many elements of the project that no one had properly examined.

White House lawyers, for example, were uneasy about the implications that would follow from declaring ownership, control, or any form of sovereign rights over this new territory beyond earth's atmosphere. Around the world, several serious diplomatic incidents had been caused by foreign planes flying over another country's land without permission. Satellites passing over enemy territory could bring those disputes to a whole new level.

Eisenhower had proposed his Open Skies policy at the Geneva Summit in 1955, where the Big Four—Britain, France, the Soviet Union, and the United States—had met to discuss ways to end the Cold War—including disarmament and the reunification of Germany. The Open Skies proposal would have allowed "mutual aerial observation."

In short, Eisenhower offered the Soviet Union the chance to see blueprints of every military installation in America and Europe and to send "peaceful" planes to take aerial photographs to verify the information. In exchange he wanted the same courtesy from his Cold War enemies. The Soviets rejected the proposal.

To keep hope alive of resurrecting disarmament talks, Project Vanguard had to be presented as a civilian project focused entirely on the advancement of science. It was run by the US Navy with direct oversight from the Pentagon, yet it had to appear nonthreatening.

If the Americans could get into space first, and do so with an object that was clearly scientific in its goals, perhaps they could persuade the Russians to approach its space program in the same spirit. Then everyone could avoid the horror of living under orbiting nuclear missiles.

For this plan to work, however, the science on board this first satellite had to be legitimate, and convincingly so. A panel was formed to decide the experiments that could be conducted from orbit. The general remit would be to observe the planet and find out more about conditions beyond Earth's atmosphere, communicating back to Earth using a telemetry system—the same type of radio transmitters being used to control missiles.

One of the winning ideas came from Maurice Dubin of the Geophysics Research Directorate at the US Air Force research center in Cambridge, Massachusetts. He proposed designing a system to measure how often Vanguard would be hit by tiny meteorites. Dubin was given a grant of $89,045 and told to create a system so sensitive that it could measure particles as small as one one-thousandth of a millimeter in diameter.

By November 1956, there had already been some test launches for the rockets that would carry Vanguard into space. Yet Dubin appeared to have difficulty getting his meteorite counter beyond an initial concept. There had been a long-running battle between the navy and its contractor, Glenn L. Martin Corporation, over the specification of the Vanguard satellite. Eventually it was decreed that it would be a very small sphere, weighing about 3.25 pounds (1.5 kilograms), with a diameter of just 6.4 inches (16.3 centimeters). Given the space constraints, Dubin was struggling with his design. There were a handful of other experiments that also had to fit into Vanguard's casing.

With deadlines looming, Dubin called Buck asking for help. It does not seem that the two men knew each other previously, although they could have met through any one of dozens of military science conferences.

Dubin appears to have believed that Buck could build a cryotron-based system to count meteorites. Buck explained that his cryotron was not ready for such a job. He could, in theory, build a computer using vacuum tubes, transistors, or even magnetic cores that would perform the task. There was no prospect that such a computer could ever be small enough to fit inside a six-inch ball, however.

Buck put forward an alternative solution. There was no need for a computer to fulfill the task, he posited. A watch mechanism could count the meteorite strikes; it was reliable, small, and unlikely to be affected by any extreme temperature swings experienced when leaving the atmosphere.

Buck's notebooks show the design he sketched out after Dubin's call. An electric sensor could be used to detect the impacts, relying on the theory of piezoelectricity; the basic principle is that certain substances give off a tiny electrical charge when a force is applied to their surface. The small electrical signal from the sensor could then be used to nudge the watch mechanism one notch every time there was a collision.

Over the following months, Buck designed the system with the aid of Ken Shoulders. It was just another experiment that started to appear in Buck's notebook. A particular entry from January 1957 records that he and Shoulders spent a good deal of time making miniature solenoids for the system. Although Buck and Shoulders do not appear to have been credited with aiding Dubin, the official description of the experiment that went into orbit in early 1958 is a match with the drawings in Buck's notebook.

Quietly, Buck and Shoulders had played their part in the very early phase of the space race. Their work is still in space, so far as anyone knows. Vanguard 1 beamed back information to Earth about meteorite collisions until NASA lost contact with it in 1964. It is not expected to drop out of orbit until 2198, however.

This minor supporting role in Vanguard 1 was by no means Buck's only contribution to the emerging space race. Another much more secretive satellite plan had begun, one that was only fully declassified in 1995 under the orders of President Bill Clinton. The WS117L program was conceived by the Surprise Attack Panel, chaired by MIT president James Killian—the same body that commissioned the U-2 spy plane.

While a high-altitude plane could be used to fly over suspected missile sites to take pictures, a satellite could do so even more discreetly. If it was in permanent orbit, spying on Earth, the cache of information would be unprecedented—it could peek behind the iron curtain completely unnoticed.

Buck's close friend Louis Ridenour, the chief scientist at Lockheed Missile Systems, was heavily involved in the program, having been unofficially handed a contract to build and launch the satellite in March 1956. The project was given the official green light in October of that year. Buck took several meetings with regard to the research; the code name WS117L is sprinkled through his diary entries at this time, with little additional information.

Lockheed's interest in Buck was escalating, not only due to this satellite program. The rockets that would propel satellites into orbit were the same as those on the missiles that could carry a nuclear warhead.

15

THE POST-SPUTNIK EFFECT

UDLEY BUCK BURST THROUGH THE DOOR OF THE LAB AT MIT, flustered and excited, with bags under his eyes. He had been up all night, but this time it wasn't because his baby son Douglas had been crying too much. He had spent the night glued to his ham radio gear, listening to an eerie stream of bleeping noises coming from space.

It was Saturday, October 5, 1957. The night before, the Russians had launched Sputnik 1, the first ever artificial satellite. It was a polished metal sphere, weighing about eighty kilograms and trailing three long, spring-loaded aerials that were pulsing a signal back to Earth. With Sputnik traveling at about eighteen thousand miles per hour, each orbit of Earth took only ninety-six minutes.

"At MIT, everyone I knew was stunned and wildly excited," recalls Bert Korkegaard, a physics student in Buck's lab at the time who had been a radar operator in the Korean War. "I remember Dudley showing up excited and exhausted after listening to its radio signals on ham radio gear. Somehow that helped make us realize it was really happening."

The Soviets had drawn first blood in the battle to conquer space. Not only had they succeeded in getting something into orbit, but it was communicating with Earth. If they could do that, they could launch a missile—or suspend one in orbit. Indeed, the rocket that had been used to launch Sputnik had already been through successful tests for use as an intercontinental ballistic missile.

Sputnik generated its own propaganda. For twenty-two days it beamed radio signals that were picked up by ham operators all over the world. Anyone with a telescope, or even a pair of binoculars, was able to see the shiny ball pass overhead.

Four weeks later, the Russians bettered themselves, launching Sputnik 2. It contained Laika, a part-terrier, part-husky stray picked from the streets of Moscow, which became the first (and only) dog to orbit Earth. She died within a few hours of launch, as she was always destined to eventually: going into space was still a one-way ticket, with no means for reentering Earth's atmosphere having yet been devised.

President Dwight D. Eisenhower tried to keep a cool reaction in public to the Sputnik launches, claiming that he knew all about the Soviet program thanks to information gleaned from U-2 missions over the Soviet Union. Nothing could have been further from the truth, however, according to NASA historian Roger Launius, who maintains that while America was still better placed, both technologically and militarily, Sputnik was a phenomenal coup in the propaganda wars.

On the night of the launch America's satellite experts had been at a cocktail reception at the Soviet embassy in Washington, DC. Scientists from either side of the iron curtain had come together for a six-day conference on space under the auspices of the Comité Spécial de l'Année Géophysique Internationale—a neutral international organization that had been created to oversee the informal competition to get a satellite into space first.

According to their official schedules, both the Americans and the Russians were on track for their first launches to come early the following year. It was a reporter from the *New York Times* attending the party who broke the news of Sputnik to the American delegation. His editor had called the embassy to say that the launch had been announced by TASS, the Soviet news agency.

Sputnik had already orbited Earth twice, without any of the US systems detecting it. Suddenly Soviet premier Nikita Khrushchev's boasts that the USSR was the greatest country on Earth were starting to carry more weight. The Project Vanguard team had been comfortably beaten. It would be another six months before Vanguard 1 was launched successfully. The first attempt saw the rocket exploding in flames about three feet off the ground; the next got to an altitude of about four miles before breaking apart.

THE CRYOTRON FILES | 169

In the end, Vanguard was not even America's first satellite to launch. Eager to put something into space, Wernher von Braun and his team sent Explorer 1, a modified rocket, into orbit on January 31, 1958—just to prove that they knew how to do it.

The political impact of Sputnik was overwhelming. Eisenhower was under pressure. He had come to power on the back of his military credentials following discontent with the Korean War. Now he was viewed as a lazy president who spent too much time on the golf course. Sputnik had cemented the idea of the "missile gap" in the public debate, and the Democrats were out to cash in.

"The only appropriate characterization that begins to capture the mood on 5 October involves use of the word hysteria," wrote NASA's Launius. "Almost immediately, two phrases entered the American lexicon to define time, 'pre-Sputnik' and 'post-Sputnik.'"

Lyndon B. Johnson, the Texan Democrat and future president, was the Senate majority leader at the time. He immediately launched hearings of the Senate Armed Services Committee to review America's space and defense programs. The committee concluded that America's space efforts had been woefully underfunded and pinned the blame directly on the president and the Republican Party.

"The simple fact is that we can no longer consider the Russians to be behind us in technology," said George Reedy, one of Johnson's aides who would serve as his press secretary once they eventually reached the White House. "It took them four years to catch up to our atomic bomb and nine months to catch up to our hydrogen bomb. Now we are trying to catch up to their satellite."

Eisenhower had started to scramble for ideas before the Johnson camp went on the attack. Four days after Laika was blasted into space, he called MIT for help. James Killian, the MIT president who had helped devise the U-2 (and wrote the blueprint for finding future scientists in *Life* magazine), was called into government full-time as Eisenhower's special assistant for science and technology.

He also instructed the US Department of Defense to set up a new technological research agency, the Advanced Research Projects Agency, to coordinate the science being developed in different branches of the military.

After a degree of badgering from Killian, his newly created President's Science Advisory Committee recommended that a new civilian agency be created to handle America's space programs—the National Aeronautics and Space Administration, or NASA.

In the first days after the first Sputnik launch, Buck's diaries and notebooks go quiet. The usual routine of entries about lab experiments, developments in projects being run by his students, and progress in finding key materials simply stopped.

He appears to have been diverted to a greater cause. When the satellite first went up, many of America's top scientists were told to drop their routine work and turn their attention to Sputnik—calculating the orbit of the tiny metal sphere and trying to predict how long it would take before it started to lose height and then burn up in the atmosphere.

As the "hysteria" that Launius described evolved into a more pragmatic response, the need for America to raise its scientific game became a national priority. Better computers were a big part of the agenda, as computers were clearly going to underpin all technological progress. They could run the calculations to put satellites in the air and they could guide missiles; they could also be used for their original purpose of cracking codes and processing intelligence.

The quicker the computer, the more it could do. The cryotron was seen as the answer—certainly in some quarters. The National Security Agency was still the dominant force in the American hierarchy when it came to all things computers.

As David Brock from the Computer History Museum explains,

There were two things going on at this time. The NSA wants the biggest, fastest, most-powerful computers it can get. They don't care if they are huge. They don't care if they need a special building. They don't care if you have to freeze it until it's colder than the outer reaches of space. They don't care how much it costs.

On the other hand, there is all the stuff going on in aerospace: ICBMs [intercontinental ballistic missiles], supersonic jets. The computerization of aerospace introduces another set of pressures. In aerospace, they were saying we care that you can shake it and it doesn't break. We care how much it weighs. It actually has to be superreliable, because if we put it in a satellite or an ICBM we can't change the part. It has to be heat resistant. It has to fly. Smaller, more reliable, more rugged. This makes everyone interested in microelectronics and microcircuitry. The cryotron was ticking off many of these boxes. The refrigeration was the only thing.

At this point, nobody really expected the silicon integrated circuit—what we would broadly refer to as the microchip—to evolve into meaningful technology. It was only in about 1964, five years after Buck's death, that silicon-based semiconductors started to take over from the cryotron.

"What first attracted me to the Cryotron and to Dudley Buck was that there was a time when he had a considerable lead," explains Brock. "These integrated thin film cryotron arrays were ahead of where silicon integrated circuits were. More complex cryotron devices had been made. They were more integrated than silicon integrated circuits."

A lot of the most enthusiastic research into Buck's cryogenic switch was being undertaken by IBM. The company had been working on different variations of the cryotron ever since it had first encountered the device in the summer of 1955.

After Buck refused their invitations to join the company, IBM set its own researchers to the task of turning the cryotron into a revolutionary computer component. One such researcher was a former University of Washington college classmate, James Crowe, who worked for IBM's Military Products Division and had developed a memory system using cryotrons that could switch in ten nanoseconds, about one hundred times faster than the magnetic core memories that were installed in IBM's commercially available machines at the time.

Crowe came to MIT and showed Buck an even quicker switch that could flip from zero to one in eight nanoseconds. It was a secretive project; Crowe waited ten years to file the patent on his discovery.

In October 1956 a full IBM cryotron research program was set up under the physicist Richard Garwin, who had joined from the army's renowned lab at Los Alamos, New Mexico. It was Garwin who had refined the design of the Ivy Mike hydrogen bomb. Soon he had a team of researchers running experiments at the company's Watson research lab at Columbia University, another at the IBM headquarters in Pough-keepsie, New York, and two other teams in temporary research facilities that had been set up across New York State.

By that time IBM had won the contract to build an upgraded version of the Semi-Automatic Ground Environment (SAGE) air defense system—the project that had evolved from the Whirlwind computer. Garwin had been seconded to the Whirlwind program during 1953–54 while he was still on the military payroll. He would have encountered Buck then, had they not met previously. IBM had been set a tough specification by the air force for the SAGE system, and believed that cryotrons could be the key to building a computer capable of the task.

Garwin spoke about the cryotron research project in a 1986 interview on his career with the American Institute of Physics: "I had a hundred people working for me at various IBM locations by the end of 1956, to build superconducting computers out of thin film cryotrons."

Buck had filed the patent for an upgraded high-speed cryotron in January 1957; three more variations on the design were filed in the following three months. By the time of the Sputnik launch in October 1957, the device was becoming more sophisticated.

The day after Killian was appointed to the Eisenhower government in response to Sputnik, Buck gave a lecture on his invention to the American Institute of Electrical Engineers in New York. It spawned yet more research projects.

A few days later he gave a more advanced version of his speech to the NSA at Arlington Hall, the former wartime codebreaking station. Solomon Kullback, the agency's director of research and development, was so keen for his staff to hear about the cryotron that he opened up Buck's part of an all-day seminar to "all agency personnel," according to a memo in Buck's files.

Building a supercomputer was now a top priority for the NSA.

An earlier scheme that had been more or less forgotten about was given fresh attention. Project Lightning had been dreamed up in 1956 by the then head of the NSA, General Ralph Canine, who wanted to chalk up one last milestone before retiring.

At a cocktail party in July of that year he had blurted out, "Build me a thousand megacycle machine! I'll get the money!" The machine he demanded was about ten times quicker than anything in existence at the time.

Eisenhower signed off on the budget for Project Lightning personally. Buck had been loosely involved in the project from the start, having been summoned to see Canine two days after he got the budget signed off.

There were a number of concerns about Project Lightning, however. As a concept, it sounded rather like Project Nomad, the giant costly flop that Buck had been sent to monitor. Both machines were conceived as a means to process large volumes of data, such as the intelligence that came flooding in from NSA listening posts. An expansion of intelligence-gathering had created an overwhelming bureaucratic burden.

Declassified NSA documents reveal the problem: "Sites around the world were sending [redacted] intercepts to NSA each month in the 1950s; conventional machines were not equal to the task of sorting, standardizing and routing this tonnage."

Post-Sputnik, Project Lightning got a new injection of energy and interest. The belief was that "with an adequate budget and a genuine 'free hand' NSA could create a new generation of super-fast computers, perhaps tripling processing speed at a stroke."

Project Lightning was not about building a single machine but instead devising the components that could allow lots of new machines to be built. The reinvigorated second phase of the project was mostly about cryotrons. IBM, which was by this stage dedicating a lot of resource to Buck's gadget and its own incarnations of the device, was one of the main contractors hired to work on the scheme.

IBM historians have previously declared that by 1958 roughly 85 percent of the Project Lightning funding was being directed to cryotron technology.

As Brock explains,

Project Lightning was the NSA saying, "Let's find the thing that's better than the transistor that will give us more gigantic computers that are faster and use less power." Lightning was a huge part of getting cryotrons going. The main points of interest were superconducting electronics and the cryotron. And a device called tunnel diodes, which burned pretty brightly then died out.

The cryotron effort in Lightning, a lot of the money, went to IBM; a lot went to RCA. A lot of other companies saw this and jumped in—like GE—even though they did not get any Lightning money. Lightning was an important shot in the arm to this microcircuitry area.

There was now so much work being conducted on the cryotron that concerns were emerging about the stocks of helium being consumed by the experiments. When big companies wrote to Buck to inquire about the cryotron, he would ask that they contact their congressman to alert him to the need for conserving helium in America's public interest.

Although there were now huge numbers of scientists researching the cryotron, the most advanced work was still taking place at MIT. It was not just the device itself that was groundbreaking, but the way in which it was made.

Efforts to build quicker and smaller cryotrons had moved in a new direction, technically. Buck and his team were trying to draw cryotrons by firing beams of electrons, using a device similar to the equipment you would find inside an old television set.

The electron guns inside a cathode ray TV set are used to excite the phosphor on the back of the screen to create a picture. In Buck's experiments, the fine beam of electrons fired by the gun was being used to start a chemical reaction. In the areas where the gun was fired, it would leave behind pathways of superconducting metals that could be arranged in different patterns to replicate the effect of winding the two tiny wires around one another.

The chemicals were deposited first on thin layers of quartz or silicon. They could be stacked on top of one another. Repeating the process with different chemicals, layer upon layer, allowed Buck and his lab colleagues to create an incredibly small version of the cryotron. As with the earlier experiments, the smaller it could be made, the quicker it should become.

The ideas they were using were not wholly original. The German scientist Gottfried Möllenstedt at the University of Tübingen had used electron guns to change the properties of chemical films. His work was focused on optics technology, however.

Using an electron gun to make a computer circuit was a new concept. It is ultimately how silicon-based integrated circuits—the microchip—later came into being. Chuck Crawford, Buck's former lab assistant, believes it was Ken Shoulders, his lab partner, who read into Möllenstedt's work and decided it could be translated for their purposes. It was Buck who then pulled the concept apart, breaking down what must have seemed a rather fanciful notion into a practical series of experiments.

Shoulders became renowned in later years for his wacky ideas. He had a contract with the CIA to build jet packs and primitive drones. He built a flying car, the Girodyne Convertiplane, but no one would let him test it on the roads. The association with Shoulders, Brock believes, has been bad for Buck's reputation over the decades since his death. The broad sweep of history has relayed the narrative of how the silicon microchip changed the world. Those with a hand in its development have been heaped with praise. Buck's work has been mostly forgotten or ignored by everyone other than those who continued the research. Yet it was Buck and Shoulders who led the pack, in some regards.

In more recent years, thanks partly to Professor Karl Berggren of MIT's Department of Electrical Engineering, the role of Buck and Shoulders in the evolution of computer chips has started to be a little more widely acknowledged. Since the mid-2000s, Berggren has been teaching MIT students about Dudley Buck and the work he did on campus. Berggren refers to all forms of superconducting chips as nanocryotrons "in Buck's honor."

Berggren also insists that credit is properly apportioned to Buck and Shoulders for devising the process of lithography—the writing of circuits with electron beams. Over a period of decades the origin of the concept had been attributed to Nobel Laureate Richard Feynman. Berggren has set essay questions where he asks students to read Feynman's paper and the Buck–Shoulders research and then identify whose vision reflects how the technology evolved. "The 'correct' answer is for the students to observe that Buck's vision was the one followed," says Berggren. He wrote an editorial for the journal *Nanoscale* in 2011 to make this point specifically:

> It has become a cliché to reference Richard Feynman's *There's Plenty of Room at the Bottom* lecture in nearly every commentary published on nanoscale engineering and fabrication. However, less well known, but perhaps more accurately visionary, was a paper written by Buck and Shoulders in 1958, a year before Feynman's speech, which laid out a procedure by which nanostructured electronics might be written by electron beams. Although their vision was narrower, it was closer to the path that the nanotechnological revolution actually followed in the ensuing 50 years.

AS BUCK AND Shoulders worked on their new electron-gun cryotron, it seems evident that they were far from certain it would work. While working on the Cryotron Mark 2, using this cutting-edge electron beam technology, they continued to run a parallel set of experiments based on improved versions of the original wound-wire design.

Their attempts to manufacture the new thin-film cryotron, as they called it, required some new equipment. The electrons were concentrated down to a beam a fraction of a millimeter in diameter using two condensing lenses. Buck and Shoulders were working on such a small scale by this stage that they did not have a microscope powerful enough to see what they were doing. Buck sunk a considerable sum of the lab's money into buying an electron microscope from RCA—a necessary tool for the experiments he planned to do, but something no one at MIT had seen before. It could blow up images to sizes one thousand times greater than standard microscopes.

"Much to everybody's amazement, not the least of which mine, Dudley said to me, 'Chuck, you are in charge of running this microscope,'" explains Crawford. "I was an undergraduate, and this fancy new electron microscope is put in my charge. The service technician from RCA was appalled. It was like he was a car salesman who had just sold a man a new car who then took his five-year-old kid, put him in the driver's seat, and said, 'Okay, you're the driver.'"

Soon Crawford was running experiments with a variety of chemicals in every possible combination to try to take the thin-film cryotron from concept to reality. The trick was to find chemicals that could be successfully converted into fine lines of superconducting metal after being blasted with a beam of electrons.

Crawford reflects on his work at this time as being similar to the research performed by the early pioneers of photography. Just as photographic film reacts to light to leave an image behind, so the different chemicals reacted to the electrons.

"For a century and a half, the human race has made silver halide–type compounds to make photographic film," explains Crawford. "That silver halide process has to be pretty carefully prepared to make it work. There were decades of experimentation learning to make first black and white pictures, then color pictures, then Polaroid instant pictures before digital photography took over. For every reaction you can make go by light optics, there are probably thousands of reactions you can make by electron optics. There's a broad spectrum of things you can do with electron optics. To figure out which reactions were worthwhile, you had to begin learning organic chemistry."

For a group of electronic engineers and physicists, this was straying into less familiar territory. They did not know too much about the finer points of chemistry.

"Dudley decided he had to understand organic chemistry better to make some of these things work," Crawford recalls. "To get the chemistry, Dudley started sitting in on organic chemistry courses. If you are a member of the staff at MIT you can audit any courses you like, so he started sitting in on organic chemistry courses."

The experiments went on endlessly. Writing circuits with an electron gun was not as easy in practice as it sounded. There were lots of botched attempts.

Other scientists around America were also starting to find ways to make smaller and smaller computer components—most notably,

Jack Kilby at Texas Instruments. By the summer of 1958, Kilby was trying to make all the components of a computer circuit on one lump of either germanium or silicon. It was an attempt to miniaturize the transistor, the revolutionary switch that had been devised by Bell Laboratories in 1947. Kilby had not yet worked out how to insulate the components, which Buck was doing by layering circuits on top of one another.

No one at MIT thought these semiconductor computer components would take off—especially not Buck and Ken Shoulders. As Crawford recalls,

> There were a couple of papers around explaining that you could never make really small transistors. The reason given was that power dissipation in the semiconductor would be high to the point where if you tried to pack a large number of them into a small space the heat would be overwhelming. That was because the quality of the silicon semiconductors was pretty lousy. It was an industry that took a number of years to come up to speed.
>
> At that point in time, for the group at MIT, making the computer components in solid state and thin layers was almost an obvious idea. It was apparently blocked by what was then regarded to be valid physics calculations that suggested it would be very hard to do.

By the spring of 1959, Robert Noyce, an MIT alumnus who was one of the cofounders of Fairchild Semiconductor, came up with an idea similar to Buck's, only using ultraviolet light particles rather than electrons to write the circuit. Noyce later cofounded Intel Corporation, one of the world's biggest manufacturers of microchips.

Buck knew that he was part of a generation of scientists that would leave a mark on history. One of these competing teams of scientists would soon make a monumental breakthrough. He suspected he would get there first. Buck's chips were superconducting, whereas Kilby and Noyce were using semiconductors. The process of making them was similar. Where Kilby and Noyce had an advantage was that their transistor-based invention did not need to be suspended in liquid helium in order to work.

In December 1957, more than a year before Noyce conceived his version of the microchip, Buck wrote in his notes, "I feel that there is a revolution in digital computer fabrication available to us in the next decade and that our present work with cryotrons and other vacuum-deposited computer components ranks high among the possible ways in which that revolution will come about."

16
A RECIPROCAL ARRANGEMENT

WHILE THE SOVIET UNION BASKED IN THE SUCCESS OF SPUTNIK and continued test-launching its intercontinental ballistic missiles, it was still significantly uncompetitive in the field of computing.

The buzz of competing ideas in both universities and the private sector that characterized the American effort to build bigger and better computers did not exist on the other side of the iron curtain. The Soviets mostly built machines to solve specific problems. They had one that was attempting to predict the weather, and another that was dedicated to attempting to crack the problems of machine translation.

The Soviets had long suspected that America had the edge on electronics. They had seen how the F-86 fighter had outmaneuvered the MiG-15 in the Korean War. Reluctant to believe that American engineering could be superior, the Russians had concluded that its superiority must come from its significantly more advanced on-board instrumentation.

The upper echelons of the Soviet science community knew about America's digital computers. The general commentary on computing advances in newspapers and magazines would have been sufficient to comprehend the technology gap that had evolved.

They wanted to know more. Since Khrushchev had come to power, bringing an "intellectual thaw" that had run through the Soviet science community, academics were permitted to have greater access to research and ideas generated outside the Soviet Union and its Eastern bloc neighbors.

In the autumn of 1957 a group of Soviet computer scientists applied for visas to travel to the United States in December of that year. They asked to attend the Eastern Computer Conference in Washington, DC—one of the prime forums for discussing cutting-edge computer technology. Surprisingly, the US Department of State thought it would be a good idea to allow them to attend, on the condition that a corresponding exchange be arranged with the Soviet Union. In anticipation of a trip to Moscow, a small group of American computer scientists was assembled by the Institute of Radio Engineers and warned that they should be on alert to travel at extremely short notice. But no invitation came from the USSR in time for the exchange to be arranged ahead of the conference, and the whole thing was dropped.

Nonetheless, a few days after the conference, the National Joint Computer Committee—a body formed from representatives of the trade organizations for electrical engineers and radio experts and the recently created Association for Computer Machinery—held a meeting to discuss an exchange program. They voted in favor of inviting a Soviet delegation to attend the same conference the following year, in December 1958, when it was due to be held in Philadelphia. In addition, they proposed to take the Russians on a tour of American computer factories and research labs. Naturally, their thinking was that the Russians would extend the same courtesy to American scientists. It was an offer of information exchange, similar in concept to the Open Skies proposal that Eisenhower had put forward two years earlier.

Mort Astrahan, a top executive at IBM who chaired the National Joint Computer Committee, was tasked with leading the negotiations. Astrahan was intimately involved in the development of the SAGE air defense system and subsequently worked on the first IBM computers controlled by a typewriter-style device. To discuss the idea of a Soviet information exchange, Astrahan contacted the State Department for help, where he was put in touch with the East–West Contacts staff, who started to advise on the logistics of such a trip.

Astrahan was told that it would be up to the committee, as hosts of the Russians, to make travel arrangements, find guides and interpreters, and get clearances from all the installations that would be visited. The committee would also be responsible for ensuring the delegation did not stray into any closed areas.

The State Department also furnished Astrahan with a long list of all the things the committee should ask to see in the USSR in return. Armed with this information, the committee drafted a letter to the Soviet Academy of Sciences in April 1958 suggesting the exchange. They received no response.

In the meantime, an academic at the University of Michigan successfully hosted a group of four Soviet academics for four days at its annual conference on digital computers in June 1958. The trip was directly reciprocated, with four academics from the university making a four-day trip to the USSR in August; they visited a number of the Soviet computer factories and research institutes and lectured on the work being done in America. Although they were on opposite sides of the Cold War, the trips were conducted in the spirit of academic cooperation.

While in the Soviet Union, John Carr, the leader of the trip, broached the subject of a bigger exchange with Sergey Lebedev, the vice president of the Soviet Academy of Sciences and the nation's top computer expert. Having already seen the benefits of an exchange, Lebedev was keen on the idea. The invitation was re-sent, but this time cabled directly to Lebedev, and followed up with a registered letter.

By October, Astrahan received a response—though he had some difficulty understanding it. Not only was it in Russian, but the Russian characters used to write the message had been garbled in the process of sending the cable. Astrahan needed help from the State Department just to get a sensible translation. The message was a long list of demands for access to American facilities, particularly the IBM installations working on the 704 machine and the newer 709. They also wanted the full program of the Eastern Computer Conference. There was no mention of a return invitation.

Given that it was already October by the time the Russians responded, organizing anything in time for the December conference was going to prove impossible. The committee decided instead to arrange a tailor-made trip for the following spring, arranging visits to almost all of the facilities the Russians had asked to see. After fourteen telegrams and four more formal letters, the trip was confirmed and its itinerary agreed upon.

Included on the itinerary was a trip to Boston to see, among others, the young assistant professor at MIT whom the Russians believed was building the guidance system for America's intercontinental ballistic missile. They had seen from the documents they had been sent that Dudley Buck was one of the expected headline speakers lined up for the conference in Philadelphia.

The conference went forward without the exchange, and Buck was indeed a star performer. He gave several talks during the event, which ran from December 3 to December 5, 1958, covering his work with self-organizing systems and trends across the nascent computer industry.

Yet the main paper he presented was the one that really left a mark. Titled "An Approach to Microminiature Printed Systems" the paper, authored jointly with Ken Shoulders, explained in detail how to make a microchip. It explained the work he and Shoulders had been doing in the lab with the electron gun to make the new, ultrafast, thin-film cryotrons.

Shoulders had just been lured to a new position at Stanford University to work in its famous research institute. He and Buck continued to work together on developing the cryotron, but it became a long-distance relationship.

Stanford was extremely keen to bring Buck to California, too. John G. Linvill wanted him to take over as the university's head of electrical engineering. It would have been quite the promotion. Buck had turned down offers of staff jobs from the University of Washington and the University of California, but Linvill was offering the chance to run an entire department and to be reunited with Shoulders.

Buck took Jackie with him to visit Linvill. They even picked out where they would like to build a house—the job came with a free lot in the middle of what is now Silicon Valley. Yet the new academic term was already under way by the time Stanford made its approach, so Buck took the view that there was no need to leap to an immediate decision on whether to accept.

The Stanford job was appealing. At first glance, it seemed a more supportive place. Shoulders published a paper through the Stanford Research Institute in October 1958, not long after he got there, outlining the work they had been doing with electron beams and thin films of metals. The day after it was published, Buck gave a talk to assorted senior professors at MIT, and the patent experts from Research Corporation, the agents that commercialized the institution's patents. They viewed it as a "blue sky" presentation, and seem to have been unsure as to what Buck had actually achieved.

Perhaps it was indicative of the cultural differences between America's conservative east coast and the unbridled optimism that characterizes California in particular. Yet it appears that there was an intellectual elite within MIT that was never quite willing to accept Buck's achievements.

There seems to have been a view at MIT that the hype surrounding the cryotron was overblown. Perhaps the litany of highly-decorated professors with their well-documented wartime achievements were reluctant to have their place in the spotlight taken by a young experimentalist who was still lacking in formal academic achievements.

Even Gordon Brown, the outspoken Australian who headed the electrical engineering department, and who had previously agitated for Buck's work to be properly recognized, appears to have been of this view. The Franklin Institute in Philadelphia wrote to MIT in November 1957 asking for a specimen example of the gadget to form the core of a planned cryotron exhibit. Brown eventually turned down the request after ignoring the museum's correspondence for as long as possible.

Thanks to his prizes and media attention, however, the outside world continued to see Buck's cryotron as a seminal invention. McGraw-Hill, the educational publisher, sent Buck a check for thirty-five dollars in July 1958 asking him to write about it for their latest encyclopedia of science and technology.

By August 1958 the US Department of Defense had taken delivery of the prototype cryotron computer that had been built for them by Arthur D. Little. So far as Buck was aware, it was not yet being used for any specific military purpose. Yet it contained "1,800 cryotrons, which, themselves, occupy 10 cubic inches," wrote Buck in response

186 | IAIN DEY & DOUGLAS BUCK

to one of his students. "The control equipment and Dewar vessel occupy 70 cubic feet."

The Arthur D. Little computer had been built using the wire-wound cryotrons. Buck, Chuck Crawford, and Ken Shoulders were rigidly focused on perfecting the new iteration of the design, produced through electron beams. Thanks to the powerful microscope they were using, and the tiny electron beams, they could see the potential to make cryotrons that were more than one thousand times smaller. They could potentially make circuits with connections that were just 0.1 micron in width—that's 0.001 millimeters. The process was detailed in an undated paper in Buck's files titled "High Resolution Etching of Evaporated metal Films, K. R. Shoulders, D. A. Buck."

The manufacturing process was still far from perfect. One of the issues was dealing with the vapors created as a result of the chemical reactions they incited with the electron gun. Some of the chemicals they were using required that a vapor linger for a few seconds to allow the reaction to occur fully. Sometimes these gasses just contaminated the whole experiment.

Buck, Shoulders, and Crawford had understood the basics of making the microchip, however, as the paper explained: "By repeating the process of depositing a metal film, selectively depositing a resist, and then etching, one can form a multilayer structure of narrow conducting paths. Between metallic layers, the same deposition process which is used to selectively form the resist can be used to deposit an insulating layer."

There was one fundamental difference about the nature of Buck's work relative to many of the other computing pioneers of the time. Kilby and Noyce were both working on their potential microchips for private enterprise, guarding news of progress in the name of commercial confidentiality.

Buck, the altruistic academic, was sharing details of his work with anyone who would listen. As a famed scientist of the age, he was in constant demand to speak at conferences or to demonstrate the technology to corporations or government departments.

There was still a problem, however, that had already been illustrated by the Arthur D. Little design. No matter how small Buck made his cryotrons, he was reliant on suspending the devices in liquid helium, and traveled around the country with his helium-filled dewar flask—carrying the highly explosive apparatus on commercial flights, trains, and the passenger seat of his car—to demonstrate his latest experiments. Although helium itself is extremely inert, if the connections got clogged with ice, as they were prone to do, the whole thing could blow up.

The Philadelphia computing conference in December 1958 was a milestone moment, however. It was there that Buck announced to the world in granular detail the process that he and Shoulders had devised.

After some deliberation, Shoulders decided against flying from Stanford to attend. Until that point, most of the presentations on manufacturing thin-film cryotrons using an electron beam had been conducted jointly by the two men. Buck was on his own this time.

"The day is rapidly drawing near when digital computers will no longer be made by assembling thousands of individually manufactured parts into plug-in assemblies and then completing their interconnection with back-panel wiring," he told the conference. "Instead an entire computer or a large part of a computer probably will be made in a single process."

The paper explained how to deposit chemicals on a surface to form a thin metal film, how to insulate different metals from one another using quartz, and how to repeat the process layer upon layer to make a complete circuit. Buck explained how, in theory, it could be possible to squeeze in fifty million components per square inch, and that it was conceivable to stack ten thousand layers of these components. There was a caveat, however:

> This degree of microminiaturization, however, would be difficult to justify in the manufacture of machines as simple as present-day computers. If a computer could be reduced to the size of a cigar box by means of one of the techniques mentioned, there is little point in reducing it further to the size of a postage stamp. The computer would already be dwarfed by its own terminal equipment and, in the case of cryotron circuitry, its Dewar vessel.

Man has ambitious plans for information-handling machines, however, and one can easily envisage a time, a decade from now, when vast numbers of components will be needed in a single machine.

Buck explained how the techniques he was developing could also be adapted to store large documents or to process information. "Such a system might be useful in the telemetering of space-probe photographs back to earth," he noted. He then talked through a specific experiment that had been conducted by Chuck Crawford in the MIT lab, using films of Parlodion, silicon monoxide, and molybdenum, with masks made from copper mesh. It explained how he let tetraethoxysilane into the vacuum system before firing the electron gun and then added chlorine to make a second chemical reaction. He explained, in other words, how he and his team had already made microchips in their lab experiments. His audience was blown away.

"Zounds!" wrote Ken Shoulders in a letter to Buck a few days after the conference, adding,

> From some of the reports I have received on the E.J.C.C. you must have had the after burners on. Some of the fellows came back here days later with their eyes still dilated—to describe your description of the future.
>
> From a purely business point of view, thanks—this may be the kind of nudge that will start our effort here down a fairly nice road. From a personal standpoint, however, it just makes me damn proud to be your associate. Keep your standards high.

Three days after the conference ended, Buck received a letter from Solomon Kullback at the NSA. It was an invitation to join a new panel being assembled by Buck's friend Louis Ridenour to advise President Eisenhower on electronics and data processing.

As a prelude to taking on this role, Buck was asked to attend a two-day seminar in Washington the following week, related primarily to Project Lightning, the advanced supercomputer program based on his cryotron.

There would be only seven other people on the panel: one from

IBM, one from Lockheed, one from General Analysis Corporation, one from Datamatic, another doctor from MIT who worked in the university's Lincoln Lab, and two from Remington Rand, the company building the Universal Automatic Computer (UNIVAC).

There was little written down on Buck's letter of invitation to explain what the panel would be about, but the work of its members suggests it clearly tied into the need to build ultrafast computers to support the space race, missile technology, and the continuing work of codebreaking.

After a decade working for the government on secret computer projects as both a full-time agent at Communications Supplemental Activities–Washington and then as a part-time consultant, Buck was being brought deeper into the inner sanctum, with an indirect line to the president. He was, in effect, on the front lines of the Cold War.

17

THE MISSILE MEN

D
R. LOUIS RIDENOUR'S RESEARCHERS AT LOCKHEED MISSILE
Systems were determined to hire Dudley Buck. Their repeated
letters made clear that they remained of the view that cryotrons
would form the basis of the guidance system for America's nuclear
deterrent.

Lockheed was already developing the Polaris submarine-launched
missiles, and the X-7 test rockets. It had just developed the Agena
upper-stage rocket—the first multipurpose spacecraft, which would
go on to be used in the Gemini space missions and which continued
to be used by NASA until the mid-1980s. It had been built, at this
point, as part of Project Corona—an offshoot of the WS117L program, of which Buck was a part.

The satellite and missile programs were more or less part of the
same work stream. Lockheed was working on both. The same rockets
that were being developed to carry a nuclear payload would be used
to launch satellites.

After his refusal of a full-time position, Lockheed tried to hire
Buck as a consultant. There was a problem, however. When his contract with the NSA had been renewed in 1956, a new clause had been
added. The NSA had started to get worried about conflicts of interest
emerging between work done by its various advisers on behalf of the
government and similar work done by contractors.

Buck's consulting work had been a source of confusion in the past.
The MIT patent department had been concerned to learn how much
work Buck had been doing on the cryotron during his time on Arthur

D. Little's premises. To maintain full patent control, the university sought assurances that the key work been completed in Building 10 at MIT, rather than the Arthur D. Little labs.

Buck had been forced to fill in a form for the NSA detailing his various consultancy agreements. Where he had once gathered consultancy contracts here, there, and everywhere, he now started to think a little more carefully about how he divided his time. Arthur D. Little also got a little more possessive.

When Lockheed sent Buck a contract, offering him a consultancy role on its missile project, he sent the paperwork back. His contract with Arthur D. Little "precludes the possibility of my signing your agreement," he wrote. He did come up with an elegant solution, however. "If your firm would like to explore the possibility of using cryotrons in missile systems, it would be perfectly proper for me to participate in such an exploration if arranged through Arthur D. Little Inc.," he wrote to Lockheed in February 1957. "That is, your firm should contact Arthur D. Little with the request that they send me as a consultant if they feel that is advisable. . . . They are, as you know, a firm of consultants and I am sure that an agreement satisfactory to the interests of Lockheed could be worked out."

Buck sent another, less formal, letter to Sam Batdorf, his main correspondent at Lockheed on such matters. He explained what he had just proposed. "Work is proceeding on the cryotron," he noted. "We are still engaged in fabricating equipment to measure automatically a number of superconductor properties. In my personal opinion, I feel that it will be some time before cryotrons will be suitable for portable use in a missile system."

Lockheed was undeterred by Buck's pessimism, and Batdorf fired off a letter to Arthur D. Little requesting Buck's services at the company's missile research center in Sunnyvale, California:

The Missile Systems Division of Lockheed is interested in light and reliable airborne computing equipment for certain potential future applications. It is possible that the cryotron or some modification of it might be advantageous as compared with other proposed systems.

In an effort to ascertain whether or not this is the case, I called Dudley Buck on the telephone recently. However, because our applications are highly classified, we were not able to get very far in considering the question over the telephone. It also does not seem too feasible to process the question by exchanging letters. I, therefore, propose that Dudley come out and visit us.

Batdorf suggested that Buck spend one day with the firm, just to see if there really was any potential for the cryotron to become the guidance system he needed. Lockheed would expect Arthur D. Little to offer Buck for free initially, Batdorf said, but he would pay the firm $125 for every day subsequent to that if there was something worth pursuing.

The following month Buck flew to California. The trip was clearly a success. Three weeks after he returned, Arthur D. Little bumped Buck's consulting fee up to two hundred dollars a day. Within a few months, Lockheed scientists were sending Buck letters with questions about how to set up their helium freezers and looking for tips on where they could buy cut-price tantalum and niobium. The cryotron was now a Lockheed research project.

Cryotron research labs were popping up all over America. In December 1956, while trying to recruit Cornell University academic George Yntema to his staff, Buck bragged about the extent to which his work was being followed. "I now know of 12 engineers outside of MIT whose full-time job is to exploit cryotrons for computer circuitry," he wrote.

The research was all focusing on the idea Buck had hinted at in his award-winning paper for the Institute of Radio Engineers. Rather than using wires wrapped around one another, Buck and his acolytes across America were trying to create cryotrons by using thin films of superconducting metals. As Buck had predicted, the smaller the cryotron could be made, the quicker it would flick between zero and one.

IT WOULD BE misleading to give the impression that Buck was entirely on top of his multiple briefs. The conflicting demands on his time had left him spread extremely thin over too many projects.

Buck's father, Allen Buck, died in February 1957—he was stabbed in an alley in an attack by a jealous lover. Dudley did not even go back to California for the funeral; they had barely spoken for years.

In the lab, he was starting to get distracted. He would become so immersed in his work that he would lose track of time. Simple domestic tasks would slip his mind.

"I would have perhaps a dollar left in my jeans at the end of the week," remembers Allan Pacela, who started working in the lab in Building 10 as an undergraduate in 1956. "On the way home, Dudley would need to buy a carton of milk, but he would never remember to bring money. I would loan him a dollar, my last dollar, so he could buy milk for his kids."

Carolyn and Douglas were growing up fast. As soon as they could talk, they were both introduced to the world of science. Carolyn would get a spoon of coffee in her milk as a breakfast treat if she could correctly recall the distance from Earth to the moon or the sun. Buck would take Douglas with him to the hardware store on Saturday mornings, so that he could "learn how a cash register worked."

By the summer of 1958, Jackie was pregnant for a third time. Buck was trying to arrange his affairs so that he could see more of his children. He would try to get home to Wilmington before bedtime, then catch up on paperwork later at night.

The house on Birchwood Road was starting to feel small, especially when Buck was trying to work from home. He decided to convert the basement, which had been more or less useless up to this point; anything more than a passing rain shower caused it to flood. "Dudley used to jest that the Ipswich River originated in his basement," remembers Jackie.

Buck asked Allan Pacela, who was always in need of some extra cash, to come out to Wilmington to help him dig a new drainage trench to take the water away from the house. It didn't really work, but Pacela got a few dollars in his pocket, a hot meal, and spent the night on the Buck family sofa.

Buck then came up with a devious plan to create a false floor that could sit above the water. He set a layer of concrete blocks tipped on their sides, with gaps between them for the water to run out. He then rigged a pump that would pull away the water when required.

Once it had dried out, Buck walled off the heating furnace and hot water tank. He then tiled the whole thing to create a playroom for the kids where they would run around on wet winter days. In the other corner of the basement, Buck set up a desk and some shelves to create a home office.

It was from here that Buck finally typed up his doctorate thesis on superconductive electronic components. Although Gordon Brown, his head of department, had instructed him to complete this task by June 1957, it was some eleven months later, in May 1958, that the paper was delivered.

Buck liked the quiet, subterranean office in the family home. Over time, more and more of his files made their way into the basement. Often the curious little puzzles that would be sent his way by the NSA would be resolved at home.

His troubleshooting work for the NSA continued to be varied. He had been pulled back into air defense projects, including the Semi-Automatic Ground Environment (SAGE) system that was taking over from the Whirlwind machine in watching the skies for Soviet bombers. Diary entries also suggest he had been involved on the fringes of US Air Force Project 438L—the code name for a plan to develop a super-computer that would be able to target incoming missiles.

Buck was also developing a deeper interest in what was then referred to as either self-organizing or self-learning systems—forms of artificial intelligence.

Self-organizing systems adapt to changing environments, in response to new information. Such a system could, for example, push a missile back on course toward its target if it got caught up in the jet stream.

Buck started to correspond with Frank Rosenblatt, at Cornell University, who was pioneering the field. Rosenblatt was another of the scientists who was also developing something that looked a bit like a microchip. Together they started to share ideas on circuits that could react to rewards and punishments, aping the basic tests to define intelligence that had been performed on rats and pigeons for years.

When it came to Buck's cryotron, there were still a great many skeptics who were either unwilling to believe in the quasi-magical powers of super-conductors or failed to comprehend how the design could ever be made to work.

After one of his conference appearances, Buck received a pointed letter from E. Mendoza, who worked at the University of Padua in Italy but had also worked with the world-renowned British computer scientists at the University of Manchester, where Alan Turing had worked post-war.

Mendoza could not comprehend how Buck proposed to dissipate heat from his device. If he could make something that switched so quickly from one to zero, surely it would generate heat. If it generated heat, then it would interfere with the temperature inside his helium flask. Given that the whole concept relied on controlling temperatures at the exact point at which a metal would switch from a supercon-ducting to a resistive state, it all sounded a bit far-fetched.

"Your question is most to the point," said Buck in his reply of July 24, 1958. "Aside from the problem of building a tiny component in the first place, the problem of getting rid of the heat which results from its operation is of prime importance. We have not yet dreamed of operating superconductive switches at as high a repetition fre-quency as you mention in your letter, but have extrapolated measured values of energy dissipation for the present wirewound cryotrons to values for much smaller cryotrons which would operate at high speed."

Buck explained the math behind his research, which showed in basic terms that as it got smaller the cryotron would use exponentially less power and so create significantly less heat; "If, for example, we are successful in our present effort to reduce the size by a factor of 100, the speed will increase by a factor of 10,000. The operating cur-rent will decrease by a factor of 100 . . . and therefore the energy dis-sipated per flip of one unit will have decreased by a factor of 10 to the power of 6."

Buck had already addressed many of the possible issues. For one, he had devised a way to capture and recycle the helium in his exper-iments, answering the complaints of those who thought he would use up too much of America's supplies of the gas with his experiments. He envisaged that the cryotron could become, over time, a standalone device with a captive helium supply built in.

When it came to controlling the temperature, Buck explained to Mendoza that one of their bigger problems was stopping heat from being conducted down the cables that connected the cryotrons in their vat of helium with the components outside the flask: "We have, for example, a student-designed cryotron computer under construction which requires 32 fine copper wires plus 16 shielded copper wires for its connection to the peripheral equipment. The heat loss due to thermal conduction evaporates one liter of liquid helium in about ten hours."

Buck said that there were countless ways around these problems and the heat problems "will probably not be objectionable in a finished system." He added, "Our present work is aimed at the problem of constructing cryotrons on a very small scale. We plan to use electron beams to replace light in an etched-wiring process which promises to allow evaporated thin films to be cut up into the desired conducting paths. A multilayer technique, then, would allow large groups of cryotrons to be made simultaneously. This work is in its infancy, however, and no progress can be reported at this time."

Many of these issues had already been addressed by the time Buck gave his well-received presentation to the Eastern Computer Conference in December 1958. Other issues were being chalked off the list day by day, week by week and month by month. Buck would make the cryotron work, if only he could find the time.

18

THE RUSSIANS HAVE LANDED

O N NEW YEAR'S EVE 1958, WHILE MOST OF HIS STUDENTS WERE preparing to go out partying, Dudley Buck was in his usual spot in the lab in Building 10 at MIT, fiddling around with combinations of chemicals. His wife, Jackie, six months pregnant with their third child, was waiting for him at home.

Project Lightning, the government supercomputer scheme that was developing cryotrons, was gathering steam. Demands for newer, faster computers to fuel the space race and the other needs of the Cold War were coming from all directions, including from the top echelons of government. At the time, the core assumption of the US military and the major computer manufacturers was that this new generation of computers would be built using cryotrons.

That New Year's Eve, Philip Cheney, an executive from Lockheed, wanted to meet Buck. Cheney worked for Lockheed's self-organizing systems division—the euphemistic technical name for the department tasked with working out how to direct a missile back on track after it had been buffeted by the wind.

Buck was spending almost all his time in the lab anyway. The theory behind his invention was now proven; he just had to resolve the practicalities of manufacturing.

On New Year's Morning 1959 his lab books and diary entries show that he made tantalum pentachloride for the first time, a key step necessary to make his microchip cryotron work. It was something of a breakthrough.

On the other side of the country, his old lab partner was also putting in hours over the holidays. On January 2, 1959, Ken Shoulders wrote to Buck from Stanford University looking for help. Shoulders wanted to secure new government funding for Stanford's work in microminiaturization and was pitching various ideas to different agencies. He was trying to strike while memories of Buck's presentation in Philadelphia were still fresh in the minds of those who mattered in Washington. He had a problem, however: the chemicals they needed for the experiments were extremely difficult to come by. Shoulders wanted to know if Buck could "spare a small sample of triphenysilalol." In a postscript to his letter, he then added, "Say hi to Jackie for me and tell her that we lost the little one. We were expecting in June."

Stanford was still trying to reunite Buck and Shoulders. The university continued to attempt to lure Buck to take the job as head of the electrical engineering department. John G. Linvill, who was leading the charm offensive, was more of a kindred spirit than most of Buck's MIT colleagues. He was also an inventor, and a veteran of Bell Laboratories. Linvill later found fame by inventing a device that could convert ordinary text into Braille; he designed it for his blind daughter, Candy.

Buck had promised to come out to Stanford in April to discuss the position further. Linvill was pushing him to come sooner, but Buck fended him off.

Stanford was not the only place offering him a post as head of department, however. He was now being pursued aggressively by the University of Kansas too.

Buck's fame was leading him in strange directions. On the same day he received the letter from Shoulders, the US Junior Chamber of Commerce wrote to tell him he had been nominated for an award. It was a slightly odd nomination, given that the cryotron was not yet being produced commercially.

Cryotron research was becoming something of an industry in itself, however. In January 1959, Arthur D. Little, the firm that had begun the cryotron research in earnest with Buck as a consultant, wrote to say that it had agreed to a "standardization of terminology for superconductive electronics" after liaising with a body referred to as Committee 28. One of the key things it did was to pin down a clean definition of a cryotron for all those using the device.

MIT was still a little wary of Buck, and his growing army of fans around the world. A handful of the more commercially orientated individuals on campus were starting to worry about the extent to which Buck was broadcasting his progress to the world and distributing copies of his papers.

On April 22, 1959, Malcolm Stevens from the MIT patent department wrote to David Black at Research Corporation, the company that commercialized the university's work, to warn him to be alert:

Dear Dave,

Do you remember the "blue sky" talk by Prof. Dudley A. Buck during your Institute visits of October 2 1958, when he spoke about the electron beam writing of circuits that could serve as memory components? It appears now that he has brought this technique much closer to reality.

Prof. E. W. Fletcher points out that at the present time the minimum dimensions for which adequate resolutions may be obtained for printed circuits and the like are in the neighborhood of two or three mils. His group is now thinking in terms of resolutions in the neighborhood of 2000 Angstrom units, which is more than three orders of magnitude superior to the present practice. One of the methods for achieving this result is described in the enclosed Buck-Shoulders paper, and on pages 6 and 7 of the enclosed Quarterly Progress Report no. 4 of the High Speed Computer System Research Program.

Prof Fletcher believes that they have a two or three year lead over some of the larger companies working in this field. The Buck-Shoulders paper or informal talks with industry may soon have them thinking along similar lines.

Your informal advice on this development would be appreciated.

Sincerely
Mal

The letter was immediately acknowledged by Research Corporation, although it would take months for the company to get second and third opinions from various legal experts and MIT bigwigs as to whether the technique of writing circuits itself was patentable.

There was good reason for concern. Dudley Buck and Ken Shoulders's paper was already spreading like wildfire through the computer industry. And others were catching up. Jean Hoerni, an engineer working with Gordon Moore and Robert Noyce at Fairchild Semiconductor, had set out some principles in January 1959 for creating an integrated circuit with what he called a planar transistor. Hoerni was a theoretician who tended not to bring his ideas to fruition; it would be the autumn of 1960 before Noyce's development of Hoerni's principle turned into an operational chip.

While Buck remained immersed in his work, family life continued to take a back seat. David, his second son, was born on March 14, 1959. As with the birth of Douglas, it seems that Buck was barely away from the lab. Entries from his notebooks on the day after the birth are full of experiments on further combinations of chemicals and their reactions to electron beams. A few days after that, the MIT graduate committee approved Chuck Crawford's proposed thesis on "effects of fine focused electron beam," allowing work on the cryotron to continue apace.

There was still a steady stream of people wanting to talk to Buck about other projects. Frank Rosenblatt from Cornell University—the expert in self-organizing systems—became a regular fixture in Buck's diary. According to sources who worked in the lab, they were working on a system to guide navy torpedoes.

Interestingly, the navy bureaucracy was ignorant of this work. Although he had avoided active service in the Korean War, Buck was still a member of the US Navy Reserve. While working on all of these multitudes of projects for various branches of the military, Buck received a letter from the Naval Officer Disposition Board telling him that it planned to honorably discharge him. He had failed to qualify for the next higher rank, partly because he had not completed a navy correspondence course on basic electronics. He had also failed to show up for Naval Security Group drills for the past three years.

Clearly hurt, Buck fired off an angry response demanding that he be retained. As an assistant professor of electrical engineering at MIT, he thought he could be forgiven for missing out on the navy's correspondence course. As for missing security drills, that was even more offensive: "I have maintained a close working relationship with my former duty station and with naval Security Group Activities. For the past six years I have been a consultant (expert, when actually employed) with the National Security Agency, Washington D.C., making visits to work on security problems approximately once per month. I feel myself to be in an excellent state of readiness to participate . . . in the event of an emergency."

This letter arrived while Buck was in the midst of work on the WS117L satellite program, elements of which have never been declassified. In early 1958, WS117L was divided into three main projects: Discoverer, Midas, and Sentry. All three projects were related to surveillance satellites of some kind or other.

The third of these programs, Sentry, was the largest and absorbed the most funds. Its budget soared from $10 million in 1958 to $159.5 million in 1959, according to a 2009 article in *The Space Review*. Yet Sentry was seen as only deliverable in the longer term.

Midas was an early warning satellite designed to detect heat signals from Soviet missiles. Discoverer, meanwhile, was the first part of the program that would be launched. Although it was officially billed as an engineering project, Discoverer was actually a cover for a covert program named Project Corona.

It was managed by the same joint CIA and Air Force task force that had successfully overseen the creation and launch of the U-2 spyplane, and under the same project manager: Richard Bissell, the CIA Deputy Director of Plans. Yet the majority of the work was being conducted by Lockheed Missile Systems, where Louis Ridenour was chief scientist. It was a "black" project for the intelligence community, meaning that it was not permitted to even confirm its existence. Corona was declassified under the Clinton administration, leading to some interesting revelations. Much of the surveillance intelligence that had previously been credited to the U-2 had been gathered instead by Corona satellites. Lockheed had been unofficially working on it since March 1956.

One of the key problems with using a satellite for reconnaissance was recovering the photographs. With the U-2, the plane would come back to Earth and then the film could be unloaded and processed as it would with any other camera. If the camera was on a satellite that was to remain in orbit, the pictures had to be recovered somehow.

The images captured by Corona would be too big and detailed to beam back to Earth via the telemetry radio connection, it was decided; there would be too much data to beam back. Although Ridenour was an expert in telemetry, thanks to his old television business, other branches of the WS117L program continued to work on this problem.

Yet to get Corona up and running, it was concluded that the film would have to be recovered from space somehow. Thus, a ludicrously complex plan was devised: they would release cartons of film after they had been exposed in a container strong enough to reenter Earth's atmosphere. As it dropped back toward Earth with the aid of a parachute, it would then be caught in midair by a plane employing a large fork-type device.

Bizarre as it may sound, the scheme eventually worked. There were dozens of failed launches, and multiple problems with recovering the film. The shiny capsules of film falling to Earth were also responsible for a number of UFO sightings—and sparked diplomatic incidents when they landed in the wrong place. One was found by Venezuelan farmers, who stripped it of precious metals before handing it over to the country's army. The American government then had to pay to get it back. Another fell in the Arctic Circle, prompting a rush to recover it before the Russians found the film—a scenario that later provided the inspiration for Alistair MacLean's thriller *Ice Station Zebra*. Nonetheless, Project Corona was ultimately a huge success.

Eastman Kodak was contracted to develop the film for the satellite. Itek, a start-up formed by former Eastman Kodak employees who had bought out a research lab from Boston University, won the tender to build the camera after submitting an extremely high-tech design. As the launch date neared, both of these companies spent an increasing amount of time with Dudley Buck, with appointment after appointment from individuals at both companies noted in his diary.

Itek won the contract for Corona after designing an ingenious system that would allow the camera inside the satellite to remain pointed at specific targets on Earth, even if the satellite itself was rotating.

In January 1959, a week before the first Corona launch, Norm Taylor, an MIT alumnus who had joined Itek, called Buck to ask for some help. Taylor wanted to know whether Buck's process of writing with electron beams could possibly be used to store photographic images. What they discussed was, in essence, whether they could build a digital camera.

Eastman Kodak, which had won the contract to design the film for the Corona satellites, soon had the same idea. Shortly after the second Corona launch (both of which were failures), Buck started receiving panicky, unsolicited messages from the company's research and development department.

"Some of our people recently attended the Eastern Joint Computer Conference where they heard a presentation by you of some remarkable results in the field of miniaturization," wrote Eastman Kodak's M. G. Harrison on April 1, 1959. "I am working on the information capacity of storage media, in particular photographic films, and would like an opportunity to talk about your work from the storage point of view."

While Buck was always keen to help, his interest was in creating the first functioning microchip, not in developing this other strand to his work that was nothing more than a theory at the time. After Buck tried to delay a meeting, Harrison started pushing quite aggressively to see him as soon as possible.

"Several weeks ago I wrote to ask for a time when we could discuss your ideas on microminiaturization in relation to information storage," Harrison wrote in his next letter. "Since this subject has generated considerable interest here I would very much like to have an early interview with you if possible."

Corona was in trouble by the time of Harrison's second letter. It would be another year before they successfully launched one of the satellites and got it to maintain orbit and to release an exposed film at the correct point.

Eventually Buck relented, agreeing to meet Harrison that June. By that date, however, Buck was dead.

MORT ASTRAHAN LOOKED at his watch nervously. He had been waiting for hours in the terminal at Idlewild Airport in New York City. The delegation of Soviet scientists he was about to show around America's top computing installations was running late. They had been due to arrive the day before, on Saturday, April 18, 1959, but they hadn't shown up. There had been no phone call or telegram to warn that they were no longer coming. When the plane showed up without them, a check of the passenger records showed that they had missed a connecting KLM flight in Amsterdam. Neither Astrahan nor his contacts at the US Department of State knew whether the group was still coming.

The following morning, however, the IBM executive got word that the group of scientists was on a flight. Their inbound plane from Moscow had been late the previous day, which was why they had missed their connection. The trip was still on. So Astrahan returned to the airport for a second day of waiting.

Astrahan, age thirty-five, worked in the Advanced Systems Development Division at IBM, based in San Jose, California. He had been building computers for IBM since 1949, starting out with the IBM 701. He had acquired a modicum of fame while working on the Semi-Automatic Ground Environment (SAGE) air defense system, having devised one of the first computers to be operated by a typewriter—the prototype for the computer keyboard.

Astrahan had started to learn a little bit of Russian, and claimed to his peers that he was able to translate, unassisted, the last of the fourteen cables from Moscow arranging the trip. Nonetheless, he had brought a translator with him to the airport who would shadow the Soviet group throughout the trip.

Eugene Zaitzeff, the appointed translator, had been made available by Bendix Systems, another computing giant of the day that would go on to design and build many of the instruments for the Apollo moon landings. Zaitzeff was computer literate, but was not a groundbreaking scientist like many of the others who joined the delegation for the Soviet visit. He was there because of his excellent command of Russian. Zaitzeff and Astrahan would later coauthor an article for the Association for Computing Machinery's Communications of the ACM that detailed the whole trip.

A second translator had also been included in the group. Professor Lipman Bers, a Latvian-born mathematician from New York University, was also waiting in the terminal, along with his wife. Bers had also been asked to act as translator for the corresponding American trip to the Soviet Union, the details of which had yet to be arranged. He and his wife had volunteered to act as hosts to the group of scientists while they stayed in New York, showing them a slice of what the Big Apple had to offer.

Astrahan, Zaitzeff, Prof Bers, and his wife waited together in the terminal, keeping an eye out for the arrival of the next KLM flight from Amsterdam. It was well into the evening before Sergey Lebedev and his delegation finally appeared.

Although they had been expecting eight men to turn up, only seven did. Vsevolod Burtsev, one of the scientists who had been part of the delegation to the University of Michigan conference the summer before, had pulled out at the last minute.

The remaining seven scientists were an impressive group. Lebedev in particular had an incredible track record. In 1951, eight years earlier, he invented the first computer in the USSR in his lab in Kiev, Ukraine. It was named the Malaia Elektronnaia Schetnaia Mashina, which translates as "small electronic computing machine." Lebedev had produced the machine with a staff of only a dozen or so engineers.

As soon as it was operational, Lebedev started lecturing the military on how his device could be used to run calculations that would be useful in the construction of nuclear bombs or, ultimately, as a guidance system for missiles. He produced an official report stating that computers could help in everything from jet propulsion to the production of energy.

It appears to have been similar in content to the report that Jay Forrester prepared for the administration of President Harry S. Truman in 1948, or the countless speeches and lectures that Buck had made about the potential for computers to change the world.

Lebedev's report on the future power of computers was passed on to Joseph Stalin. The Man of Steel then requested that Lebedev be placed in charge of a new computing institute that would be tasked with building a bigger computer. Imaginatively, they called it the BESM, the Bolshaya Elektronnaia Schetnaia Mashina ("big electronic computing machine").

A large new building was built for Lebedev in Moscow to house his research team. New premises were a rare luxury for the head of any government program in the postwar USSR. When the building was opened, the new institute was given the full endorsement of Nikita Khrushchev, the future leader, who at the time was head of the Moscow Communist Party.

The BESM had been a success, and one that the Russians kept secret from the West for several years. The only problem was that Lebedev had not devised a way to mass-produce the machine; the prototype was the only one that existed.

Nonetheless, by the time Lebedev set foot in New York, he was already working on a bigger and better BESM II, with a staff of up to 150 dedicated to the task.

There is no doubt that Lebedev, who was fifty-seven at the time of the trip, had connections to the very top of the Kremlin. His right-hand man was Vitaliy Ditkin, assistant director of the Soviet Academy of Sciences' computing center in Moscow, a dedicated mathematician who specialized in using computers to create complex calculation tables and mathematical graphs.

Then there was Yuriy Bazilevsky, an engineer who sat on the USSR's Committee of Radio Electronics. Bazilevsky had designed several computers, including the Strela system that was being used by the Soviet Air Force for calculations related to missile defense. It had also been used to determine how missile warheads could be expected to explode on impact. His primary interest on the trip was to find out about American techniques to manufacture better computer components that could become part of his new machine, called Ural.

Bazilevsky had been on foreign trips before with Lebedev, not the least of which had been to Darmstadt, West Germany, in October 1955. That was when they had unveiled the BESM machine to the world—the first time the Soviets had acknowledged publicly that they were even developing multipurpose computers as opposed to dedicated single-task machines.

One member of the delegation, Sergey Mergelyan, had spent two months studying at the University of Rome. He was the director of the Institute of Electronic Computers in Yerevan, Armenia, where he was building three new computers at the time: two that operated from the old vacuum tube technology and one called Razdan that would become the Soviet Union's first computer to be built with transistors.

Viktor Glushkov, the fifth member of the delegation, would go on to become one of the USSR's most decorated computer scientists. A talented mathematician who had produced groundbreaking work on some of the classic problems of pure mathematics, Glushkov was a relative newcomer to the world of computers. Nonetheless, he was director of the Kiev Computing Center that had been set up by Lebedev.

Glushkov, then thirty-six, went on to win a string of state science prizes throughout the 1960s for his work on cybernetics and artificial intelligence—concepts that were being pioneered at MIT at the time of his American visit. Under Stalin's rule any such work would have been totally taboo—and life-threatening if uncovered. Glushkov was leading the Soviet charge to change this.

"The negative ideological image of cybernetics in the late Stalinist period did seriously narrow the range of the first Soviet computer applications," writes MIT professor Slava Gerovitch in his book *Newspeak to Cyberspeak*. "Soviet computing was shaped by the tension between the practical goal of building major components of modern sophisticated weapons and the ideological urge to combat alien influences."

Building machines to drive robots that could replace humans in a production line, or devising computers that would attempt to replicate the human brain, was considered by Stalin to be a capitalist evil. That was why many of the Soviet machines had been built to handle a single specific task: computers were just another machine, not an electronic brain.

Perhaps the most elusive member of the delegation was Victor Petrov, the director of the Moscow Computer Factory. Even after the two-week stay was over, Petrov's American hosts had gleaned little about his work. His factory mostly made mechanical, analog computers, he claimed, and had not yet delved far into the digital world. "Very little information on his factory could be obtained from him," wrote Zaitzeff after the visit.

The last member of the group was Vladimir Polin, an engineer who worked directly for Lebedev. He was a molecular physicist by background, but was now tasked with building a translation machine. He appears to have been the most junior member of the delegation.

When the delegation arrived in the airport terminal after their long flight, the seven appear to have been in a fairly foul mood. Although they exchanged friendly handshakes, the trip had taken its toll. Their hosts bundled the group into their cars and whisked them into town to their Manhattan hotel. Astrahan and Zaitzeff, both out-of-towners themselves, were staying in the same hotel as their Soviet guests.

As soon as they had checked in, Lebedev got down to business. The Soviet computer guru made it clear that he was not entirely happy with the itinerary that had been planned for them. The key problem, he explained, was that he felt they needed more time at the IBM plant. "He stated that in his country changes of schedules such as this could be very easily and swiftly arranged," wrote Zaitzeff.

To the Russians, IBM was something of a mystery. Soviet computer technology had been kept firmly inside the military. All the machines that the delegation had designed and built were being used to calculate ways to battle America. While IBM had more than its fair share of military work, an increasing amount of its equipment was being used to run payrolls for businesses across America, work out tax calculations, or crunch numbers on census data.

Mort Astrahan, conscious that the itinerary had been agreed through extensive diplomatic coordination, flapped a little. He said he would "try his best" to fit in a second tour of his employer's factories toward the end of the two-week stay, and suggested that they might be able to replace one of the other visits that had been planned with a second trip to IBM. It was something of a hollow promise, but it seemed to go down well.

"Although the visitors must have been very tired, they then invited their hosts to a vodka and caviar party in one of their hotel rooms," wrote Zaitzeff. "After friendly conversation a few arrangements were made for the last two days' stay of the group in New York, which would include dinner at Professor Bers' home, and also a Saturday trip through the stores on Fifth Avenue conducted by Mrs. Bers. Then the group broke up and each retired to his room."

That Monday morning was wet and miserable. After breakfast, the first stop the group had to make was to the Park Avenue building where Soviet diplomats to the United Nations were based. They were given bankers' drafts—their allowance for the two weeks they were in the United States—and told to cash them at the First National City Bank.

They went to the nearest branch, but were told that the drafts they had been given could only be cashed at the bank's head office on Wall Street. Already the trip was taking on a farcical tone.

As Zaitzeff wrote,

Since the group was expected in [IBM headquarters at] Pough-keepsie by 1:30 that day, time was running out. We decided to check out of the hotel and pick up the cars which were rented for the trip to Poughkeepsie. Somehow, in the confusion, the first cab took off without any English speaking individuals inside. Dr. Astrahan and myself found ourselves with the rest of the group in the second cab. Thus, upon arrival at the Hertz office and pay-ing off the cabs we found that Mr. Glushkov forgot his briefcase inside the cab, not knowing that this was his final stop, and thinking that the cabs would be used to take us all the way to Poughkeepsie. Everyone became even more concerned when we were told that it contained three bottles of vodka and about six pounds of black caviar. It was never found.

In the rented cars the group was driven to Wall Street to get their cash. It was noon by the time the two cars got on the road, leaving just an hour and a half for the trip. Even by twenty-first-century stan-dards, they were cutting it close.

"After being separated on the Taconic Parkway, we each thought the other was ahead and were almost arrested for speeding," wrote Zaitzeff. "We were reunited by a flat tire, just as the police closed in, and all was well!"

IBM had lined up some Russian speakers from among their ranks to join the welcome party. Some of them were wartime emigrants from Eastern Europe. After a quick round of introductions in a conference room, the delegation was led on a tour of the plant. The first thing they were shown was an IBM 705—the machine that was being installed in offices across America. Then they were taken to the third floor, where they saw the production of magnetic core memories—the technology that Buck had helped invent, leading to the legal battle between IBM and MIT. Afterward, they were brought right through the main production line for the IBM 704 and 705 machines just as the day's work came to an end. Then they were shown IBM's state-of-the-art transistor factory.

Of all the questions peppered by the Russians, the theme that struck their American hosts as the most telling was their concern with the reliability of the IBM machines. They were particularly intrigued by the magnetic tape that was used by these computers. How could information be reliably written onto these tapes, they asked?

After dinner Astrahan took the group to meet some of his friends from IBM, presumably to give them a flavor of ordinary family life for an American computer scientist. Then they went to their modest motel, "which, by the way, they liked more than any of the hotels they stayed in later," according to Zaitzeff.

The next morning saw more presentations and more demonstrations, interspersed with coffee breaks. The Soviets also handed over some books and papers that they had brought for the IBM team, detailing some of their work.

They were shown an early attempt at artificial intelligence: an IBM 704 machine was running an experimental program for playing checkers. Ditkin, Lebedev's number two man, decided to take up the challenge: the machine conceded after only a handful of moves.

A lecture on machine learning followed, where Glushkov led the questioning. Most of the rest of the delegation were still obsessed with asking about the reliability of the IBM machines, apparently incredulous at the fact these computers could run for hours on end without breaking down.

Under cross-examination the Russians then started to reveal a little more about their own work. They were not developing machines that could play board games, they said; their only dabbling in machine learning was strictly related to improving industrial processes. To translate languages, they believed that a dedicated special-purpose system was required; they were still just formulating the specification for such a machine. As part of this, the Russians said, they were devising a new system of abbreviations that could help reduce the amount of memory this new computer would need.

This must have sounded strange to the Americans, who were already dedicated converts to the belief that all progress in computing depended upon faster and bigger memories. Reducing the complexity of a task to make it easier for a computer to handle was already an outdated way to look at the problem.

The Russians then started to talk about their weather forecasting computer, named Pogoda, which translates simply as "weather." It could do one-month forecasts or twenty-four-hour forecasts, they said, and it took about a half hour to run the calculations.

When the IBM executives started to ask for a little more information on which data points they used to produce these forecasts, "Academician Lebedev stated that he did not know" and shut down the conversation.

By the end of the day Lebedev received some good news. The State Department and IBM had agreed to his schedule change. They could come back to IBM on their last Friday before returning to New York.

In exchange, Mergelyan assured the Americans he would let them come to Yerevan to see inside his Razdan machine when they made their corresponding trip. There had been no plan to take the group on the long trip to the Lesser Caucasus region of the Soviet Union before then.

It was a relief for the Russians to hear that they would get to come back. They had only just started to scratch the surface of one of their main themes of interest: the cryogenic computers that IBM was making; they knew they could find out more about that on the next leg of their trip.

At 4:30 p.m. the group was driven to LaGuardia Airport to get a flight to Boston. The next day they would meet Dudley Buck.

It is only logical to assume that Soviet intelligence had a sizable file on Buck. His involvement in attempts to convert Konrad Zuse to the American cause back in 1948 would have been known; Zuse was hot property at the time. Further, from at least the summer of 1952 Buck was being approached by agents from Amtorg, the Soviet import-export agency in the United States. This was a fact known to both Allen Dulles, the US director of central intelligence, and his predecessor, Walter Bedell Smith. By the time the Soviet delegation had arrived in the United States, Dudley Buck was an extremely high-profile computer guru.

Given that Khrushchev had instructed his rocket scientists to read every technical report they could procure on American missile developments, one has to assume that Buck's widely distributed paper on the cryotron had made its way into the hands of the KGB. And given the games that were being played by the likes of Louis Ridenour—who believed in letting the enemy know bits about what was going on in America to sustain the bleak logic of mutually assured destruction—it is only logical to assume that the Russian delegation knew all about it.

The seven Soviet computer scientists were eager to meet Buck. They sat at one of the big round tables in the MIT faculty hall, waiting for lunch. They had spent most of the morning in a conference room being shown diagrams of how MIT organized its departments and research programs.

After the Russians handed over some of the documents they had brought with them detailing their work, Frank Verzuh, one of the MIT computer experts who had worked on Project Whirlwind, took them on a tour of the university's IBM 704 installation.

In the afternoon they were scheduled to see MIT's famous cryogenic computer lab—home to the tiny computer that *Life* magazine had proclaimed two years earlier could become the guidance system for America's long-range nuclear missiles.

Buck strolled into the room and introduced himself. He had brought with him Norbert Wiener, the world-renowned mathematician and expert in cybernetics. Mergelyan, the Armenian who was building the USSR's first computer based on transistors, had met Wiener before on a trip to India. Wiener believed that automation of industry could alleviate poverty by creating more efficient economies, and he was an official adviser to the Indian government at the time.

Wiener's views were the polar opposite of what Stalin had believed, but they would soon gain credence in the Soviet Union, mostly thanks to scientists who were on the trip.

Mergelyan's English was good enough for the two men to chat without Zaitzeff acting as interpreter. The Armenian computer expert was so excited by what he was hearing that he didn't touch any of his lunch.

For Eugene Zaitzeff, the translator from Bendix Systems, it was a rare opportunity to finish a meal. "I usually had to time-share eating with being an information flow channel," he wrote.

After lunch the group was led up to the third floor of Building 10 and into Buck's lab. The room was a hive of activity, as usual, but no one was making cryotrons or trying to build microchips.

Buck told them that the helium canisters needed to operate the cryotron were being refilled and thus a demonstration was impossible. Given that the visit had been in the diary for months, it seems unlikely that the Russians would have believed this excuse. Ever patriotic, Buck clearly just didn't want to show them his work.

"Nevertheless, a very detailed and comprehensive explanation of the work was presented," wrote Zaitzeff. "The Russian group was quite interested in the way the helium is saved in operation in the laboratory, and were surprised to learn that this is not being done. Since helium is quite cheap in the United States, there is no necessity for saving it at the present time. On the other hand, in Russia helium is not found in natural state, and thus it is imperative that it be circulated without a large loss of the gas during experiments."

Buck had already devised ways to recycle helium, and fears of a helium shortage had emerged not long after Buck started his work, when the naval dockyard in Boston shut down the free supply of the gas for MIT's work with superconductors. Abundant helium was clearly a line of propaganda that Buck was feeding his visitors.

The Soviets were soon on their way to the servomechanisms lab, where they were shown a computer-controlled milling machine. Then they were shown the TX-0 machine that had been built at MIT using transistors. It was hooked up to a cathode ray tube with a light pencil, similar to the one Buck had invented years earlier to operate the Whirlwind machine. It was programmed to play tic-tac-toe. The Russians experimented with the setup by writing ever smaller with the light pen until Whirlwind could no longer tell an X from an O.

Then they were given a lecture on the new TX-2 machine that was replacing the TX-0. They didn't get to see it, however. The group left MIT that night full of questions, not least about the cryogenic computer they had not been able to see.

They spent the next day at Harvard University, then flew to Philadelphia before spending a few days in Washington, DC—which included time for shopping. They were given a tour of the National Bureau of Standards office, and the Bureau of the Census, where a Universal Automatic Computer (UNIVAC) was processing the data. Then came the Federal Aviation Agency, and the Patent Department, both running off automated data processing systems courtesy of computers installed within the previous few years.

It was an extraordinary level of access to America's top facilities in both the public and private sector—a fact that was frequently acknowledged by the visitors. Their visit was no secret, however. Local newspapers were waiting for the group with photographers at many of their stops along the way. Most nights ended after dinner with vodka and caviar provided by the Russians for their respective American hosts.

Zaitzeff, who had stuck by the group throughout their trip, was bombarded with questions every step of the way. The group asked the population of each city they visited and were astonished when Zaitzeff told them he did not know, or gave only a rough approximation. They were baffled to see the number of skyscrapers that were being built in New York—they had been told in the Soviet Union that America had concluded this was an inefficient way to build that always lost money.

They grew a grudging respect for American cars, and asked a torrent of questions about the cost of each make and model they encountered on the road. The answers "were not readily believed when the conversation involved used car lots," wrote Zaitzeff.

They wanted to know the salaries of everyone they met—even taxi drivers. One taxi driver in Boston explained that he earned more than the average cabbie because he owned his car. The Russians did not believe him, as he had a radio in his car, "to which the taxi driver explained that five or six of the fellows got together and organized their own small cab company each owning their own automobile. That impressed the group quite a lot."

Eventually, at the end of their two-week stay, the Russians were taken back to IBM headquarters in Poughkeepsie, New York. This time only Bazilevsky, Lebedev, and Petrov went to the computer giant's plant, while Ditkin and Mergelyan gave lectures at New York University; Polin and Glushkov stayed in town to finalize the plans for their return trip.

IBM showed the group how it manufactured the magnetic cores that ran its machines, then took them to the company country club for a round of golf—which was of particular interest, given it was famously President Dwight D. Eisenhower's favorite pastime.

"Each of the visitors had a chance to use a 9-iron and putter later on the green," wrote Zaitzeff. "To the great delight of everyone present, they were quite good considering that none of them had ever held a golf club before."

That night they drove back to New York City, confident that they had seen the IBM factory "from top to bottom." Zaitzeff then took them to see a production at the Cinerama—one of the widescreen movie theaters that were the height of fashion in 1950s America, as Hollywood tried to fight back against the new threat of TV.

The Russians concluded that, technically, Cinerama was probably superior to what they had back home but that the "direction was inferior" to that with which they were accustomed.

After two days of shopping and sightseeing, and a dinner party hosted by Bers and his wife, the group was shipped back to Idlewild Airport to catch their flight.

The Russians had a colossal stockpile of intelligence to take home to Moscow. IBM clearly had nothing to hide, having opened its doors twice to the Soviet delegation. The famous young engineer at MIT who seemed to be building the Americans' missile guidance system had been nowhere near as forthcoming, however.

19

A PACKAGE

O N SUNNY WEEKEND MORNINGS THE BUCK FAMILY HAD STARTED
to take the mail boat from New Bedford, Massachusetts, out to
Cuttyhunk Island, part of the Elizabeth Islands chain that runs
south from Cape Cod. It was a quiet spot, just across the sound from
Martha's Vineyard.

Buck had plans for Cuttyhunk. He had met a Boston wool merchant named C. W. Wood at a meeting of the Navy League who had
told him about some land he owned in the area that he wanted to
sell—a miniature island right by the harbor at Cuttyhunk, with no
electricity or running water and a tiny strip of beach to itself. The
Woods invited the Bucks to come and visit them so they could take a
look. After a couple of trips, Buck decided he would build a vacation
house there and reached an informal agreement with the Woods to
buy the miniature island.

"The land was wonderful," remembers Buck's wife, Jackie. "It
overlooked the harbor, and the open ocean beyond. It was quiet and
peaceful, except for the occasional mournful sound of a distant fog
horn, and the bell of the buoy. We fell in love with the place instantly."

Buck wanted his kids to learn to sail, a skill he had mastered as a
student in Seattle. He wanted a place where they could roam free. The
idea of a house on Cuttyhunk was one of the factors that gave him
cause for hesitation about making the move to California.

He had just been given yet another honor for his work, the prestigious Eta Kappu Nu society's Outstanding Young Electrical Engineer
Award.

With the original wire-built cryotron still making waves, the microchip version nearing completion, government work flooding through his door, and some of America's top universities fighting over his services, Buck was feeling good about life. Soon, he reckoned, he would be able to fulfill his long-held ambition of owning a sailboat. He started to tell Jackie that everything was about to click into place, that their days of using old bedsheets as curtains were now far behind them. He joked about their early married life when he used to remark that it was "purifying to be poor." He invented stories about how they used to go to the store to buy a single tin of beans to share between them, just to hear Jackie laugh.

Wilmington had evolved into a nice, suburban family town, just as Buck had expected it to when he had bought the newly finished house. Their kids were growing up with a big group of friends from the neighborhood who were constantly in and out of one another's houses. The parents would take turns to babysit.

Buck was starting to talk openly about how life was about to change. He never specified how or why, but he gave Jackie the impression that money was about to start rolling in.

Earning money was already getting easier. Buck had just been offered four hundred dollars to write an article for *Scientific American* magazine. When he had started out on the Whirlwind machine nine years earlier, it would have taken him three months to earn that much.

Since the birth of David, a sturdy redheaded baby boy who had weighed in at nine pounds, seven and one-half ounces, Buck had taken his wife on a couple of dates, including one to the opera. They were planning a trip to Europe—Buck had been invited to speak at a conference in Paris that summer, all expenses paid, but they were extending the trip. He had gotten in touch with one of his many uncles, Dudley H. Buck, who was living in Italy with his wife Mabel. Uncle Dudley, who had been posted to Verona with the US Army Corps of Engineers, said he would drive up to Paris and take Dudley and Jackie back to Italy for a few days.

The family were also planning another summer driving trip across the United States, taking in Seattle, where Buck still owned the plot of land on Vashon Island that he had bought as a student and cleared with his automated stump-pulling machine. They then planned to drive all the way down the coast to Grandma Delia's place in Santa Barbara, California. Delia had not met her great-grandchildren.

The gas money for the mammoth cross-country trip was to come from the University of Michigan, which had secured Buck's services for a three-day series of lectures that he planned to give en route to the West Coast.

Buck wrote to Glenn Campbell, his former foster son, to invite him along for the trip. Glenn had just finished his two years' army service and was back in Washington, DC, lodging with Joe Keller, the Washington lawyer who had helped Buck to clear the paperwork for fostering Glenn Campbell all those years ago. Buck was worried that Glenn would go off the rails without the structure of the army. Through his role on the Wilmington, Massachusetts, school committee, he was trying to get Glenn a high-school diploma—he had failed to graduate after running back to Washington just before Dudley and Jackie's wedding. All Glenn had to do to receive his diploma was take one exam.

"I wish it were possible just to take one of the diplomas and put your name on it, but I am sure you would have none of that," wrote Buck to Glenn. "The piece of paper will have meaning for you only when it has been earned."

Buck celebrated his thirty-second birthday three days after the visit from the Soviet scientists. On the evening of his birthday, another controversial foreign dignitary came to town. Fidel Castro, who had swept into power in Cuba four months earlier, was on a publicity tour of America.

He had been invited to the country by the American Society of Newspaper Editors. With his olive-green military fatigues, shaggy beard, and tales from the field of combat he was a natural draw. In between his pledges to nationalize all industry in Cuba, he ate hamburgers and hot dogs for the cameras, went on a tour of Yankee Stadium in New York City, kissed models and babies for photographs, and took trips to the zoo.

President Dwight D. Eisenhower refused to meet the revolutionary and his entourage. Within the year, Ike had ordered the training of a

Cuban-born force that could reclaim the island nation, which would lead to the famously botched Bay of Pigs invasion two years later, once John F. Kennedy was in the White House. Castro did hold a conference with Vice President Richard Nixon, however, in what was reportedly a frosty encounter.

Although Eisenhower and Nixon had already formed their view, America as a whole was still working out what to make of the thirty-three-year-old Cuban president. The former secretary of state, Dean Acheson, the architect of the Marshall Plan for postwar reconstruction of Europe, also met Castro and subsequently described him as the "first democrat in Latin America."

The lecture tour of America was a provocative affair. In a speech to the Council on Foreign Relations in New York, Castro pledged not to accept economic assistance from the United States, and then stormed out in protest at some of the questions asked by the audience about his brutal rise to power.

Harvard University was one of the stops on the tour. Buck bought eight tickets at a price of ninety-five cents each. He and Jackie took their neighbors Nancy and Paul Bodenstein, John and Jeannie Collins, friends through the school committee, and Dick and Shirley Harding, the new minister and his wife.

Harvard held the event at the university sports stadium, but kept the capacity down to six thousand people due to security concerns. As a result, it was something of a hot ticket. It was viewed by Dudley, Jackie, and their friends as a fun night out to see the new celebrity in the world of global politics.

"It was a huge event because he was very much in the news at the time," Jackie remembers. "He was the revolutionary, and everyone was wanting to listen to him. We really didn't know the full extent of what he was up to."

Certainly no one knew that three years later Castro would become a crucial figure in the Cold War by agreeing to house launch sites for Soviet R-12 nuclear missiles with a sufficient range to hit Florida.

Four days after going to see Castro's speech, exactly a week after having brushed off his Soviet visitors, Buck was summoned again to Washington. It was a formal invitation to attend the first full meeting of the new computer advisory panel created for President Eisenhower and led by Louis Ridenour. It was to take place over two days at the new headquarters of the NSA at Fort George Meade, Maryland, on May 25 and 26. The main topic of discussion was to be Project Lightning, the key NSA program to turn Buck's cryotron into the American government's first proper supercomputer.

With missile tests going on at locations across America, the space race well and truly under way, and secret projects like the Corona spy satellites struggling to get off the ground, the pressure to perfect this new, slim-line, ultrafast computer microchip was mounting.

BUCK HATED WAITING. He would climb staircases two or three steps at a time. He never had enough time to do all the things he wanted to do. His patience was being particularly tested now that he knew he was close to finding the definitive process to create the first microchip. A whole team of undergraduate and postgraduate students was working under him in the ordered chaos of the lab in Building 10. Each had been allocated a thesis subject to test one particular aspect of Buck's work.

Each experiment came a little bit closer to success, but there were still glitches. They had found ways to make films of superconducting tantalum. Now they were perfecting the methods to layer different metals between the superconductors and write whole circuits using the electron beam. Sometimes the chemicals did not react to the electron gun as expected, and sometimes the gasses emitted during the reaction would pollute another part of the experiment. Buck had sent photographs of some of the lab results to Ken Shoulders at Stanford University to see what he made of it all.

In the meantime, the group in the lab was tinkering with all of the variables in their experiments: adjusting the temperature or altering the chemicals involved, the length of exposure to the electron beam, or the size of the current passing through the superconductors. The electron guns were replaced, then modified to try to improve the results. Different sizes and shapes of glass vacuum tubes were used in the experiments.

Buck decided he needed someone in his lab dedicated to producing custom-designed glass tubes for variations on the experiments. He found a physics student from Little Rock, Arkansas, named Jim Simpson willing to take on glassblowing alongside his other duties in the lab.

The pace of work in the lab was extreme, and the learning curve steep. Chuck Crawford ended up being awarded his bachelor's degree and his master's degree simultaneously, then got a doctorate about ten months after that—all on the back of his experiments with the electron gun.

The experimental setups got ever more complicated. Simpson, in particular, created some bizarre contraptions after he started dabbling with diborane—an extremely reactive chemical with six hydrogen atoms per molecule that explodes on contact with air. The Russians were using it as a rocket fuel, but Simpson was using it to try to create a superconducting chemical film.

Glass containers, glass stopcock valves, and various other self-created glass fittings were rigged together on his workbench behind two protective screens of heavy-duty wire mesh.

"Jim, I think I hear a tiny hiss, I think you might have an air leak," Crawford remembers saying one morning to Simpson while walking into the lab. He strolled on toward Buck's office as Simpson leaned into his experiment to see if he could work out if anything was wrong. The moment Crawford got through the office door there was an almighty explosion.

"I feared that I would see Simpson dead," recalls Crawford. "The entire apparatus was utterly demolished. Then I saw Simpson wipe his forehead, so I knew he was alive."

Two large glass stopcocks had flown within millimeters of Simpson's head, blowing through both protective screens and leaving behind two perfectly round baseball-sized holes. At the moment it all went up, Simpson had been crouching with one ear next to the apparatus, trying to locate the hiss. Although nothing had hit him, the blast had damaged his eardrum—he was never again able to hear a high-pitched noise above about three thousand hertz.

The lab was fairly comprehensively destroyed, but Buck covered it up as best he could to avoid upsetting the MIT administration—

and to ensure they could press on with their work unhindered. The formal report on the incident recorded it as a "small explosion."

The next morning Buck had to confront the quiet, conscientious little man who came each month from the Boston Edison Electric Company to read the electricity meter. There had been a "small accident" the night before, explained Buck. Without saying a word, the visitor walked to the spot on the wall where the meter should have been. There was just the grubby outline of the meter left on the wall. The rest of it was blown into a dozen pieces, scattered across the room. In the far corner, the meter reader spotted the dial. He walked across, picked it up, recorded the numbers in his book, and then said goodbye and left.

From that point on, diborane was scrubbed off Buck's list of potential chemicals that could be used to make ultrafast cryotrons. There were still plenty of chemicals that they posited could be turned into thin, superconducting films if an electron beam was fired at them, however. He was ever more determined to get to the solution.

Buck could see clearly how their experiments could change the world. At home with Jackie, he had started to talk a lot about how computers could aid the field of medicine. He was sketching out loose ideas in his basement office at home on how to design a computerized health diagnostic system. His mind was running quicker than his experimentation.

"It was the end of my senior year, April of 1959, when Dudley had the idea that we have got to get electron lithography working," remembers Crawford. "With his knowledge of organic chemistry he picked out maybe a dozen or fifteen chemicals that he thought could be catalyzed by electron bombardment to change into something else. That's what you needed."

The chemicals he needed were not all easy to come by. Without the new materials, it was difficult to do very much. Eventually, on May 18, 1959, he got word that the United Parcel Service delivery from his chemical supplier was on its way.

There was just a week to go until his first NSA summit on Project Lightning. Maybe he would find the magic formula before then.

The day dragged on as he waited for the parcel. Buck sat at his plain wooden desk, overlooking the lab, filling his day with jobs he had been putting off. First, he wrote a character reference letter for Jackie's sister Gwen, who had applied to be an assistant nurse at the nearby Foxborough State Hospital, a mental institution.

"I have known Gwendolyn Wray for five years and have employed her as a student technician in the MIT Digital Computer Laboratory during one summer," he wrote. "She is a reliable worker who follows instructions and at the same time demonstrates resourcefulness. She is a very steady worker and a pleasant team member. I would judge her personality to be well adapted to the care of mental patients, especially in situations where kindness and warmth are essential. She is, of course, a person of unquestioned loyalty."

He wrote some internal memos to other MIT professors, chasing up thesis grades for some of his students. Then there was also the usual pile of fan mail from academics and corporate titans around the world to wade through, all asking for information on his tiny supercomputer in the making.

L. G. Bishop from General Electric's defense electronics division, another of the big contractors developing cryotron technology, had written for more information "regarding the technique of writing circuits with an electron beam." Bishop was coming to Boston at the end of May and wanted to meet.

Charles H. Clark from Boeing's pilotless aircraft division in Seattle had also written, asking for a meeting to discuss the cryotron and his new method of writing circuits. (Today this is the division of Boeing that manufactures drones.) At the time, it had just installed the fleet of Bomarc antiaircraft missiles across the United States and Canada, tasked with knocking Soviet bombers out of the sky should they appear over the horizon. The missiles, which looked like ultrasleek fighter jets pointed toward the sky, were hooked up to the Semi-Automatic Ground Environment (SAGE) air defense system that Buck had worked on.

It was a remote-controlled guided missile. The SAGE radars would pick up a target, and the missile would be fired and directed remotely toward the incoming plane via a radio signal. Boeing, like its archrival Lockheed, was trying to move the technology on to the next step where the whole guidance system could be stored on the missile itself. Although Buck had been working with Lockheed on possible guidance systems for two years, he wrote back to the Boeing executive and agreed to meet him.

As Buck finished dictating the letter to his secretary, Carol Schupbach, a parcel was delivered to his office door. It was his shipment of chemicals. He shouted across to Chuck Crawford to come into his office. "He had such enthusiasm, which was characteristic of everything that he did—enthusiasm that you rarely see in anyone other than a small child going to a sports game or a movie or something like that," remembers Crawford.

The deliveryman placed the box on Buck's desk and asked for his signature. Buck started ripping into the cardboard, desperate to get his hands on his new toys.

As Crawford explains,

> He was really anticipating that one of these chemicals would be the breakthrough. They were basically bottles of slightly different sizes and shapes. Dudley just kind of looked at these things. He looked at a bottle and unscrewed the cap and looked inside—it was typically some kind of powder that we were going to evaporate on some kind of deposit and then hit electron beams on it to see if we could make anything happen.
>
> He looked at all these chemicals. He didn't eat any of them, he might have stuck a finger in a bottle. We paid no attention whatsoever to the potential hazard of taking a bunch of organic chemicals that hadn't really been studied that carefully and messing around with them. The thought hadn't entered our heads.

It was quite late by the time they had worked through the whole box. Buck started complaining that he didn't feel too well. He left Crawford to plan their next phase of experiments with the new chemicals and headed for home. George Moss, one of his other students, had just handed him the first draft of his thesis. Buck stuffed it in his briefcase, which was bulging with papers and notebooks as usual, and promised Moss he would work through the draft that evening so that they could talk about it in the morning.

Carol Schupbach was still in the office, typing up students' papers for cash; it was how she made a little extra money on the side. As she saw Buck walking out the door, she stopped him to ask a question.

"He was carrying his briefcase as if it weighed a hundred pounds," said Schupbach. "He stopped to answer my question and put the briefcase down—it was just too heavy. When I commented that he looked like he wasn't feeling well, he said, 'You cannot tell what is wrong with something until it finally breaks down.'"

JACKIE HAD NEVER known Dudley to be sick in his life. He was a man of boundless energy. That night, not long after he walked through the door, he announced that he was going to bed. It didn't look like there was much wrong with him—he didn't appear sickly in any way.

By 3:00 a.m., Buck was coughing heavily. By 5:00 a.m. he had a high fever and was barely able to move. Jackie brought him a glass of water, which he sipped slowly. "This will be the last glass of water I take in this house," he said.

Jackie called the local doctor, who ordered an ambulance immediately to take Buck to Winchester Hospital, where he had been just eight weeks earlier to pick up Jackie and baby David. When the ambulance arrived, one of the neighbors, Irene Ely, volunteered to look after the kids while Jackie went to the hospital. David started crying as soon as his mother left. Carolyn, age three, and Doug, age two, sat on the front porch while their dad was strapped onto a hospital gurney and pushed into the back of the ambulance.

"Tuesday morning Jackie called up and said that Dudley wouldn't be in—he wasn't feeling well," recalls Crawford. "In the middle of the day she called back again to say that he was quite sick, the doctors were concerned that he had been exposed to some toxic substance or something because he had been close to all these bottles. I had been just as close as he had. They wanted a list of every chemical. I prepared a list and transmitted it to the doctors."

MIT hit the panic button. Harriet Hardy, head of the university's occupational health service, was dispatched to Winchester Hospital straightaway. Morton Schwartz, an infectious disease specialist from Massachusetts General Hospital, a teaching hospital associated with Harvard University, came soon afterward, accompanied by David Rudman, professor of infectious diseases at the Harvard Medical School.

They pumped all kinds of antibiotics into his system, but nothing worked. Fearing the worst, Jackie called for Dick Harding, their local minister, who then hovered in the background while the doctors pondered the case.

The assembled doctors were desperate for another opinion. By about midnight that night they had decided the chemicals were not to blame. They concluded it must be some kind of lung condition. They had been trying to reach Max Finland from Boston City Hospital, a world expert on lung infections and pneumonia who had invented an antiserum for a virulent flu strain that plagued Boston in the 1930s. He was in Switzerland at a conference, but when they eventually tracked him down, Finland's advice was clear: give Dudley penicillin only.

"But by this time Dudley's body had been swamped by the latest high-powered antibiotics in a frantic rush to save him," remembers Jackie.

At 8:20 a.m. on May 21, 1959, a little more than forty-eight hours after he was admitted to the hospital, Dudley died.

20

THE EXTRA PIECES
OF THE PUZZLE

JACKIE BUCK WAS IN SHOCK. THE DEATH OF HER HUSBAND, Dudley, was just too much to take in. Her mother had come to look after the kids, who were far too young to understand what had happened. David, who was just eight weeks old, would never know his father. Doug's only memory of his father is that of the funny truck with the red light on top that had taken him away.

A police cordon had been set up around the house. The NSA was a little uncomfortable about what might be in Buck's basement office. Everything he was working on was sensitive, but it was possible Buck had taken home work related to the Corona spy satellites, or details of his talks with Lockheed on missile guidance systems. There would certainly be papers linked to Project Lightning, the supercomputer project based on Buck's cryotron.

MIT was also uneasy; documents crucial to future valuable patents could well be stored in Buck's basement. Gordon Brown and Ewan Fletcher, Buck's two superiors from the electrical engineering department, were dispatched to Wilmington, Massachusetts, to re-move everything even vaguely work-related from Buck's house on Birchwood Road. It was only one day after his death.

"They came and they just took all his notebooks," remembers Jackie. "I didn't have a choice. They just came into my living room, and sat down. I wasn't sure why they were there, and they said they needed all of the notebooks back and they intended to take them that day. Dudley also kept a folder full of ideas that he wanted to explore. When I went to his files a few days later, they had taken that too—or someone had taken it."

In Building 10, the record player was silent. It was Carol Schupbach, Buck's secretary, who had broken the news. After they hadn't heard anything about Buck for a couple of days, she had called the house. Jackie's mother had answered the phone. Carol's piercing scream was enough to let everyone know what had happened.

"We knew Dudley was sick, we heard he had been taken to hospital the day before," remembers Bernie Widrow, who worked in the office neighboring Buck's in the lab. "Carol Schupbach let out a scream like, 'Oh, my God almighty,' and I figured it out in an instant. Nobody expected that. Two days before he was in the lab working away. He wasn't feeling well. Everybody was telling him, 'For Christ's sake, why don't you go home, get some rest?' The next day he didn't come in. The next day he was taken to hospital, and the day after that he died."

The group of young students that had gathered around Buck was particularly hard-hit. "It was one of the saddest moments of my life, because to me he was my hero, he was a man to look up to, a man that almost substituted for my own father," remembers Allan Pacela.

It fell to a heartbroken Chuck Crawford to call Ken Shoulders out in Stanford. "I was shattered, everyone was shattered. I called Ken, and he was just speechless."

After a few days of shocked disbelief Shoulders, in particular, started to ask questions. The cause of death was officially recorded as beta-hemolytic streptococcus, a throat infection that can cause tonsillitis, rheumatic fever, and scarlet fever, and in rare cases evolves into a flesh-eating virus. The theory was that it had been lurking in his body for weeks, but that Buck had been fighting it off quietly. Yet there was never an autopsy conducted to prove or disprove the theory; at the time, there was no legal requirement to conduct a postmortem examination, and Jackie couldn't bear the thought of Buck being cut up. "I didn't do an autopsy," she recalls, adding,

He was so vibrant, so young, and I was so shocked. Looking back, he should have had a proper autopsy. I just didn't want to think about that. I was naive; I was only twenty-five when he died.

The truth is that he died so quickly after he was in the hospital that they didn't really have time to get the lab results and the blood work. That's why there was such a mystery as to why he had died. MIT was concerned he had died from something pneumonia-like in his lungs. It could have been something there that could have given him the semblance of something pneumonic. They were concerned they had not had something hooded properly, or something like that.

MIT's official recorded version of Buck's death, uncovered during the research for this book, is extremely strange. A letter sent by Hardy, the university doctor, to Ewan Fletcher claimed that Buck had become ill sometime on the weekend of May 9 to May 11 and had been off work most of that week.

Dudley and Jackie had looked after the three children of their neighbors Nancy and Paul Bodenstein that weekend, in addition to their own three. Buck had been absolutely fine. Contrary to the letter, he had not taken a single day off work until the day his package arrived. While some former students interviewed in the course of our research have claimed that he had looked ill during lectures, and that he was lacking his usual enthusiasm, that does not tally with the recollections of those who worked closest to him. Everyone from his lab said it was only late that Monday afternoon, after the package arrived, that Buck had complained of feeling ill. More important, Jackie had seen nothing wrong with him.

As the father of a newborn, working all hours of the day to finalize a groundbreaking invention while simultaneously handling dozens of classified government projects, it is understandable that Buck may have looked a little tired. But to his nearest and dearest he had not complained about anything.

An MIT faculty meeting on June 9 resolved to "place on its records its own deep sense of loss" and to pass a copy of its memorial memorandum to Jackie. It referred to his "sudden and untimely death," almost contradicting the official medical account prepared by the university's own doctor, and then went on to praise his work:

During his nine years at the Institute, Dr. Buck made outstanding contributions to the field of electrical engineering, chiefly in applications to computers of solids and their low-temperature properties.

In 1957 the Institute of Radio Engineers awarded Buck the Browder J. Thompson memorial prize for his paper on the development of the cryotron, the first miniature electronic device making use of superconductivity at extremely low temperatures. This breakthrough in low temperature superconductivity has been of great significance in the world of electronics in general and to computers in particular. During the past two years Dr. Buck had carried miniaturization even further, and at the time of his death was seeking to make cross-film cryotrons with dimensions of only a few millionths of an inch.

Buck's loyalty to the department and the "thoroughness" of his teaching were praised, as was his work on recruiting trips to high schools, noting that "during these tours he demonstrated what one observer termed a 'contagious quality of optimism, enthusiasm and just plain joy about each man's particular work.'" The memorandum continued, "Dudley Buck's originality and creativity were a constant source of wonder and admiration to his students and to his associates. He combined a warm heart with a lively intelligence of great reach, and a seriousness of purpose with the enthusiastic energy necessary for its implementation. He had all the attributes of greatness and before his untimely death had already achieved much. His loss will be keenly felt by his profession and by his colleagues at the Institute who loved and admired him."

Ken Shoulders was never satisfied with the official explanation of Buck's death. He was convinced that Buck had been poisoned by something in one of the chemicals that was used in his experiments—perhaps some of the gasses that were released by the chemical reactions. He eventually zoned in on tetraethyl orthosilicate, believing that exposure to the compound had created a condition in his lungs that looked like some form of pneumonia.

David Brock at the Computer History Museum in Mountain View,

California, believes Buck's death could have been caused by exposure to poisonous gasses. On that final day in the lab he used boron trichloride gas. The process for depositing that gas to make a test microchip releases hydrogen chloride. Exposure to either gas could cause fatal pulmonary edema, leading to symptoms similar to pneumonia.

Yet Chuck Crawford disagreed with Shoulders's view at the time. Since then he has disagreed with anyone who has posited similar suggestions of chemical poisoning.

If Buck had been killed by something from the experiments, then surely Crawford would have contracted it too. He was the one doing much of the hands-on work in the lab, and he had been fully checked out after Buck's death.

Crawford is no conspiracy theorist. He had no idea about Buck's second life with US intelligence and security services until being contacted in regard to this book. There was a long incredulous pause on the end of the phone as Buck's dealings with the NSA, his encounters with Soviet scientists, and his work on top-secret projects was explained.

For Crawford, Buck's frequent trips away had always been explained away as MIT business or science conferences, if they were explained at all. It's certainly true that the government work overlapped with his science, in any case.

The lab team in Building 10 did not know that, four days after his death, Buck was due to fly to Washington for the first meeting of Ridenour's new computer advisory group.

Is it mere coincidence that Buck died twenty-nine days after Khrushchev's top computer scientist, and six of his colleagues, were given a guided tour of the lab? That might be credible were it not for the fact that Buck was not the only elite America computer scientist to die suddenly and mysteriously that day.

LOUIS RIDENOUR CHECKED in to the Statler Hotel in Washington, three blocks north of the White House, on May 20, 1959. He had flown in from Los Angeles a few days ahead of his key summit at NSA headquarters, presumably to squeeze in some Lockheed business.

Two months earlier Ridenour had been promoted to the position of vice president of Lockheed; the company's new Electronics and Avionics Division had been separated out of the Lockheed Missile Systems business and handed to Ridenour.

A lot of the big government contracts worth winning over the subsequent few years would be directly related to the space race. While there would be rockets to build and satellites to launch, there would also be lots of money to be made from the instruments and control equipment that would be needed. Lockheed, therefore, set up a standalone division to handle this work. Having been one of Lockheed's top scientists for four years, Ridenour had wanted to prove that he could also be an executive "capable of making a profit." He was certainly close enough to government thinking on computers to know where the opportunities might lie. The work of the new division would be directly tied to his work with the NSA on supercomputers and to Project Lightning.

The morning after Ridenour checked in to the Statler he was found unconscious by a maid. He had suffered a brain hemorrhage at some point overnight. Ridenour died at almost exactly the same time as Buck. He was forty-seven, and in apparently good health.

Time magazine's June 1, 1959, issue carried obituaries of the two men on the same page:

> Died: Dudley Allen Buck, 32, exuberant M.I.T. electrical engineer and miniaturization expert, who developed the tiny cryotron to replace the transistor, was working on a cross-film cryotron (diameter: four-millionths of an inch) that would reduce a computer from room to matchbox size; of virus pneumonia; in Winchester, Mass.

> Died: Louis N. Ridenour Jr., 47, top-notch nuclear physicist who, despite being emotional about his specialty (in 1946 he wrote a grim, prophetic, one-act play about flocks of satellite bombs orbiting 800 miles above the doomed earth), pioneered in missile programs as chief scientist (1950–51) of the Air Force, helped develop the Polaris and X-17 missiles as research director of Lockheed Aircraft Corp.'s missile-systems division, became a

Lockheed vice president last March; of a brain hemorrhage; in Washington.

Nobody at the time seems to have connected the two deaths—not publicly, in any case.

The links between Buck and Ridenour were through their cloak-and-dagger dealings with the NSA. There would be no reason to assume a connection otherwise: one died from a brain hemorrhage, one from a pulmonary condition; one was in Washington, the other in Boston; one died overnight, the other over a period of days.

There have been some odd references over the years that seem to tie the deaths of the two men together, however. Howard Aiken, the American physicist who worked on the Harvard Mark I computer, claimed to have been with Ridenour on the night he died.

In a heavily redacted oral history interview for the Smithsonian Institution conducted in February 1973, shortly before his own death, Aiken also referred to a mysterious young man that appears to be Dudley Buck:

> I was very, very fond of Louis. I was with him the evening he died, in Washington. It was a very unusual thing. There was a young man who was interested in very low temperature devices in computers—what was his name? Well at any rate, he and Ridenour and I talked [redacted], and we [Aiken and Ridenour] went to Washington.
>
> I met Ridenour and we bummed around Washington all evening with him. And the last thing he said to me before he left was, "You're going to the Cosmos Club. I wish I was going to the Cosmos Club, but I can't go because I'm now Vice-President of Lockheed and I've got that two-room suite," and he shoved off and [redacted]. And the next morning, I got up at a quarter to eight and Ridenour didn't show up and this other man didn't show up. Somebody wanted to know where they were and I said, "Well don't worry about Louis, he probably has a hangover this morning." And a couple hours later, the manager of the Statler called up looking for me and telling me that Louis had died in

bed. And almost immediately after that, we got a telegram that [redacted] was dead, so we just folded the meeting and everybody went home. It didn't seem very worthwhile to go on.

It seems difficult to believe that the young man who was an expert in "very low temperature devices" could have been anyone other than Buck. He was the inventor of low-temperature devices, and was on the panel for that reason. Why his name has been redacted from this record is a curious mystery.

Sergei Khrushchev refuses to believe that his father sanctioned any kill orders on American scientists. "Assassinations and all these things were not part of the behavior at this time," he explains. "Now you can assassinate anyone, anywhere, using drones and other things. At that time, you needed exceptional things. To assassinate one of the scientists in the United States, it is such a crazy idea I don't think anyone could discuss it—not only in government, but in the KGB itself."

He is even more adamant that Lebedev, Glushkov, "or any of these individuals" who came to visit Buck would never have been complicit in any such plot. Khrushchev posits that he should know: he had encounters with all of these men while working on the Soviet missile program.

"They are not going to say, 'These are my competitors, please kill them,'" he adds. "When I was working with Russia's missiles, we tried to do our best. The Americans tried to do their best. It was beyond our imagination to think about this type of thing. Forget about this conspiracy theory. I, one hundred percent, do not believe in these things."

It is certainly true that Soviet executions of foreign agents happened more frequently in Ian Fleming novels than in real life. Nonetheless, they did occur. That notion has fed into the greatest conspiracy theory of all time.

A CIA paper on Soviet killing and abduction techniques was produced in February 1964 for the President's Commission on the Assassination of President Kennedy. The document, which runs through the repertoire of KGB assassins of that time, was declassified in 1993. It provides evidence that Soviet agents knew countless ways to dispatch targets without leaving any detectable sign of an assassination: "It has long been known that the Soviet state security service (currently the KGB) resorts to abduction and murder to combat what are considered to be actual or potential threats to the Soviet regime. These techniques, frequently designated as 'executive action' and known within the KGB as 'liquid affairs' can be and are employed abroad as well as within the borders of the USSR. They have been used against Soviet citizens, Soviet émigrés, and even foreign nationals."

The CIA knew of several assassination operations since Nikita Khrushchev's rise to power, confirming that "the present leadership of the USSR still employs this method of dealing with its enemies." The tendency to bump off rivals had not died with Joseph Stalin. The report further notes,

> The sudden disappearance or unexpected death of a person known to possess anti-Soviet convictions immediately raises the suspicion of Soviet involvement. Because it is often impossible to prove who is responsible for such incidents, Soviet intelligence is frequently blamed and is undoubtedly credited with successes it actually has not achieved. On the other hand, even in cases where the Soviet hand is obvious, investigation often produces only fragmentary information, due to the KGB ability to camouflage its trail. In addition, Soviet intelligence is doubtless involved in incidents that never become officially recognized as executive action, such as assassinations which are recorded as accidents, suicides or natural deaths.

Based on confessions from former KGB agents, the paper claimed that the Russians had been using highly advanced, untraceable poisons that could simulate death from assorted medical conditions since at least 1957, two years before Buck's death.

Bogdan Stashinsky, a former KGB assassin, admitted that he had killed the Ukrainian writer Lev Rebet in Munich with "a poison vapor gun which left the victim dead of an apparent heart attack," the CIA document shows.

He delivered the poison with a small aluminum cylinder approximately fifteen centimeters long and three centimeters in diameter that weighed about two hundred grams. It had a fine metallic screen over one end, and contained a liquid poison, hermetically sealed in a small plastic container. A small explosive charge in the other end of the cylinder could be activated by a detachable trigger that then drove a piston inside the cylinder that smashed through the capsule. As soon as the poison was exposed to the air, it vaporized. Agents were advised to use the tube at close quarters, just a few inches from the subject's face. But it was effective from a range of up to about fifty centimeters.

As the CIA report notes,

> The effect of the poison vapors is such that the arteries which feed blood to the brain become paralyzed almost immediately. Absence of blood in the brain precipitates a normal paralysis of the brain or a heart attack, as a result of which the victim dies. The victim is clinically dead within one and a half minutes after inhaling these poisonous vapors. After about five minutes, the effect of the poison wears off entirely, permitting the arteries to return to their normal condition, leaving no trace of the killing agent which precipitated the paralysis or the heart attack.
>
> Allegedly, no foreign matter can be discovered in the body or on the clothes of the victim, no matter how thorough an autopsy or examination. The liquid spray can be seen as it leaves the weapon, however, and droplets can also be seen on the face of the victim.

Stashinsky confessed to using the same type of weapon against exiled Ukrainian nationalist politician Stepan Bandera in October 1959—a death that was also officially recorded as a heart attack. Bandera was part of Reinhard Gehlen's secret spy network, the same CIA organization that Buck had been assigned to on his two trips to Germany.

The Russians had also developed a weapon that they called a "noiseless gas pistol," the Kennedy assassination inquiry papers claim. Powered by a three-hundred-volt battery, it fired a lethal, odorless, unidentified gas that was effective from a distance of up to twenty meters and took between two and three seconds to act. It was a contact poison that was absorbed through the skin and was still effective through layers of clothing.

In March 1955 the Russians used poison in an attempted abduction of Lisa Stein, an interviewer for an American propaganda radio station in West Berlin. She was fed candy doped with a poisonous drug called scopolamine. She was expected to fall ill on her way home from the café where she had met her contact. She would then have been picked up by a waiting car. She did not become ill until she was almost in her apartment, however, where neighbors were able to get her to a hospital. After forty-eight hours of severe illness, she was fed an antidote.

One other example of a poison method appears to have been included in the memorandum, although this was never declassified. Yet the CIA recognized that they did not know all of the Soviet techniques: "There appears to be no consistency in the use of poisons by Soviet intelligence to cause disability or death, or in the repetitious use of any one drug. Chemicals which have been used in cases known or suspected to be Soviet-instigated include arsenic, potassium cyanide, scopolamine, and thallium. Other likely substances are atropine, barbiturates, chloral hydrate, paraldehyde and Warfarin. Combinations of two or more substances may also be used, which further complicates diagnosis and tracing."

The KGB clearly had highly trained chemists on its books, devising ever more advanced poisons. Based on the evidence, it is certainly credible that they would have been able to devise a deadly agent that could be concealed in Buck's shipment of chemicals. Although Chuck Crawford was with him as they sorted through the chemicals, it was Buck who opened each of the assorted bottles and vials as soon as the box arrived.

It's equally possible that one of these advanced weapons was used in a more conventional manner. Buck lived a normal suburban life and worked on a big university campus full of people from all over the world. He was not penned-in behind security guards and high fences. If the KGB had wanted to get him, it could have. After all, agents from Amtorg, the Soviet export agency, had found Buck's lab seven years earlier.

It seems unlikely that the group of Russians who came to visit Buck were directly involved in his death. Yet they were clearly very senior figures who had the ear of the Kremlin.

Former KGB agents, such as Oleg Kalugin, the former bureau chief in Washington, have claimed that all Russians who came to the United States—even students—were actually spies. Of all the things the Soviet delegation heard about on its trip to the United States, Buck's inventions would have been by far the most surprising.

Soviet intelligence would already have had Buck marked as a potential inventor of the nuclear missile guidance system thanks to his appearance in *Life* magazine. Given that double agents were operating on both sides of the iron curtain, there is every chance that the KGB had been leaked a copy of the list of attendees for President Dwight D. Eisenhower's new committee on supercomputers.

At the time of Buck's death, the KGB was under the leadership of Alexander Shelepin. Countless histories of Khrushchev's reign portray Shelepin as operating somewhat outside the control of the Soviet leader. Khrushchev biographer William Taubman, for example, has described how the KGB chief ran dirty tricks campaigns against the likes of CIA chief Allen Dulles without the knowledge or consent of the Kremlin. Shelepin was later a key figure in the 1964 coup to depose Khrushchev.

The CIA paper concluded that by 1964, the time of its writing, the KGB would only resort to murder "in the case of persons considered especially dangerous to the regime." As the designer of the guidance system for America's intercontinental ballistic missile, and the architect of its most promising new computer devices, who had also been intimately involved in America's efforts to put spy satellites into orbit, Dudley Buck was arguably more dangerous to the Soviets than anyone else.

CHUCK CRAWFORD WAS supposed to spend the summer of 1959 at the Los Alamos National Laboratory in New Mexico. Buck, ever keen to see his protégés spread their wings, had given him a glowing reference months earlier to help him secure the position. He stayed until the end of July, then MIT called and begged him to come back; after

Buck's death there was no one at the university other than young Crawford who knew enough about the cryotron program.

The whole research program was severely hobbled by Buck's death. The students were passed on to other thesis advisers and allowed to continue their work, but without Buck to lead the way, the program started to drift. Those who had been closest to the charismatic young professor were given some leeway around deadlines and exams.

"After Dudley died, I continued in the lab, and the amazing thing was my job didn't go away—for reasons that I never understood," Pacela recalls. "He was gone, and I cried, I missed him, and he went away, but the job was still there and I could work in the lab, and turn in my hours, and I began to work on my own projects."

The NSA ensured that research into the cryotron continued. In 1960 Horace Mann at TRW, an electronics firm that had been involved in the construction of the Atlas and Titan missiles and went on to play lead roles in the space race, filed a patent for manufacturing cryotrons using Buck's method. He filed a related patent later that year, which was assigned to Space Technology Laboratories, the company working on the scientific payload for NASA's Pioneer spacecraft.

Mann then modified the original patent again in 1961, explaining in more detail how to make cryotron microchips.

After Buck's death, work also continued at IBM, where there was still a team of at least a hundred set up to build cryotron computers as part of Project Lightning. The team was also working with electron guns and thin layers of chemicals to build ever faster cryotrons; by 1961, two years after Buck's death, it had built a functioning forty-bit memory chip from 135 cryotrons. There were still issues with the production, however. A group was set up that changed the design, but it somehow killed the speed, with this new cryotron switching between one and zero at speeds a hundred times slower than expected.

Between 1961 and 1963, IBM worked on a project for the US Air Force to build a cryotron memory that could be used to derive associations among different types of data. That, too, ended up being scrapped.

By this time both Jack Kilby and Robert Noyce had gotten their semiconductors working. The silicon age of computing had been born. The cryotron, with its need for helium tanks and superconducting temperatures, was overtaken.

Many of the techniques used to make these new integrated circuits were the same or extremely similar to those deployed by Buck. In the case of some of the techniques, it looks like Buck could have gotten there first. The key element in Intel cofounder Noyce's integrated circuit patent was the use of an "insulating oxide." It appears that Buck had come up with this idea—and lectured about it extensively—before Noyce got to it.

Buck, however, never got around to filing his patent. Had he done so, a lot of the value in Noyce's multibillion-dollar patent would have been undermined. The earliest integrated circuit patents from Kilby, meanwhile, make no mention of this key part of the technology.

It took years for Kilby and Noyce to be credited as the joint creators of the microchip. Kilby eventually won the Nobel Prize for Physics in 2000 on the back of the invention. Noyce was dead by this time, and the Nobel Committee does not make awards posthumously. Many more have a legitimate claim for a share of the glory. Had Buck lived just a few months longer, he would have had the opportunity to advance his work to the point where his contribution would be beyond debate.

MIT knew that Buck could have shared in the commercial spoils of this discovery. From the moment the university seized his notebooks from the basement office in his home in Wilmington, it worked through everything it could find to see if there was enough to justify a patent. Eventually the case was closed, with a memo from the investigating scientists to Ewan Fletcher in November 1960, some eighteen months after Buck's death:

> I have gone over this material with Ed Thomas of the Lincoln office, who evaluated some of the Buck and Shoulders work in this area a year or two ago. The conclusions he reached at the time still appear to be valid. Namely: (1) That the scientific phenomena were known at the time Buck commenced work; (2) That Buck had laid out a

promising area of experimentation; (3) That the actual inventions upon which valuable patent coverage might be obtained would lie in the work still to be done.

While continued work by Dudley Buck might well have led to valuable inventions in this area, I am sorry to have to say that it appears very doubtful that any valuable protection could be based on the work completed before his death.

Buck was already a source of consternation for the MIT legal department by the time of his death. The university was locked in a legal battle with IBM and RCA over the patent for magnetic core memories—which eventually yielded $25 million in license payments. Any attempt to enforce unfiled patents on the cryotron, based on Buck's lab books, would have seen MIT open litigation with IBM on a new front. Buck was a central figure in the magnetic cores case.

The basic argument was that he had been too liberal in sharing information with other institutions who were part of the industrial cooperation agreement overseen by the government. A young engineer at RCA then exploited his generosity, and filed a patent based on what he knew of MIT's work. It may seem obscure now, but at the time it was one of the biggest patent disputes in history.

JAN RAJCHMAN CAME to America in 1935 as a twenty-four-year-old graduate student. He was born in London to Polish parents, but grew up mostly in Geneva, where his father worked for the League of Nations. After securing his degree in electrical engineering, he wanted to get involved in the newest technology. That meant coming to the United States.

"I was fascinated by electronics, which was the great new field at the time, and a field far more advanced in America than anywhere else," Rajchman told an interviewer for the Smithsonian Institution in 1970. "Moreover, there was the depression. It was exceedingly difficult to get a position in Europe, particularly for somebody who was not a citizen of the country where he resided, which was my case, since I was then a Polish citizen residing in Switzerland. On the other hand fortunately America had the tradition of accepting immigrants from all over the world, even though there was great depression in America, too. But still, immigrant or no immigrant, everyone was on the same footing as far as getting a job."

Rajchman did a summer course at MIT partly to dust off his English—he had not used it since the age of seven, when his family left London. In the autumn he was hired by RCA, which was designing television sets and working on radar equipment, to work in its research lab.

By the time World War II broke out, RCA was starting to dabble with primitive computing components, mostly designed by Rajchman. He was then seconded to the University of Pennsylvania to work on the ENIAC project, where he met many of America's other computing pioneers.

Rajchman never lost touch with MIT, and was regularly on campus wining and dining various professors on his RCA expense account. He became friendly with Buck and Papian. Like the MIT team, Rajchman was also working on computer memory—in league with the Institute for Advanced Study at Princeton University. He had invented a computer memory called the Selectron, another variation on the TV tube design that was being sold in RCA's machines.

It was technically brilliant, apart from the fact that it kept breaking down. The first RCA computer had seven of these tubes. According to some of the experts who used the machines, one of the seven tubes usually had to be repaired every twenty minutes or so.

Along with every other computer designer in the United States, Rajchman had been part of the information flow around using magnets to store data. He was working on an extremely similar design to that of the MIT lab. He had also built his own oven to make tiny doughnut-shaped magnets.

"During our work on the Selectron I thought of the core memory in a way," Rajchman told the Smithsonian. "But the fact is, it's hard for me to imagine the day when I hadn't thought of the core memory. I thought about it for years before writing anything down."

Rajchman filed a patent for magnetic core memory in September 1950. It was May of the following year before Forrester filed his patent—which included his idea of three-dimensional computer memory. Forrester's patent filing was based on the paper that had been sent to Eachus, but souped up with the work that Buck and Papian had done in the lab.

Both MIT and RCA continued to improve their technology, and to share ideas, as they were obliged to do under the cooperation agreements. Buck's lab notebooks record one meeting with Rajchman in October 1951 where he was shown a switch developed by the RCA lab. It was remarkably similar to something designed by Ken Olsen, Buck's close friend and colleague at MIT, who would go on to become one of the first captains of industry to emerge from America's booming new computer business as the founder of Digital Equipment Corporation.

"The switch is his own invention," wrote Buck in his lab books, of his conversation with Rajchman. "And dates to 'the weekend before my last visit' to his laboratory."

It's hard to detect from the lab book entry if Buck suspected Rajchman had ripped off the work that had been done in the MIT lab. Soon there would be no doubts in the minds of the MIT leadership, however.

Executives at RCA often wrote letters to Buck asking for details of the latest projects. As one of the big corporate supporters of the industrial cooperation agreement, it was entitled to do so.

"We are specifically interested in circuit operation and operating conditions of the Whirlwind flip-flop, gates, matrices and buffer amplifiers," wrote Lowell Bensky, an RCA executive who had previously worked in the MIT lab in one letter to Buck in December 1951.

Shortly after requesting the information from Buck's lab, Rajchman filed a second patent that included some of the developments that had been perfected by Buck and Ken Olsen. According to Bensky, this was no coincidence.

"Jan Rajchman came up under this [industrial cooperation] program and went up to MIT to talk to Forrester," explains Bensky. "Forrester had this idea for core memory. All other types of memory were subject to failure at the time. So Rajchman came up to talk to Forrester and, according to the rules of the cooperation agreement, Forrester told him everything he was doing. Rajchman then went back to RCA and wrote up a patent based on what Forrester had just told him. And the RCA lawyers went ahead and filed that patent. When MIT found out that this was what had happened, they weren't too pleased. MIT sued RCA—and they won, of course."

The legal spat was not quite as straightforward as Bensky suggests. At the time the patent scam was uncovered, nobody quite appreciated the value of the invention at the center of the dispute. During the years of argument that followed, however, the sums at stake got bigger and bigger.

Magnetic cores were clearly a better technology than the valves, tubes, and drums that had gone before. By the time the magnetic memories were commercially available, IBM had already built its first big fleet of machines using the older and less reliable vacuum tube technology; these first generation IBM machines were already installed in banks, insurance companies, and government departments on long-term lease agreements.

As Papian wrote in a memo to Forrester in July 1960, while they gathered evidence for the case,

> IBM dragged their collective feet for some time on the question of magnetic-core storage, and were not seriously damaged by such foolishness only because of their dominant business position in the field and a talent for "product-design" which enabled them to put good units into their machine lines in a hurry once they made up their minds. I recall receiving an unexpected tirade on the subject from the manager, or assistant manager, of the component development side of IBM's research division in Poughkeepsie sometime around late 1953 or early 1954. The gist of said gent's remarks was that IBM's machines were doing fine with their storage tubes, that the customers were "foolishly" demanding core storage because of our (MIT's) ridiculous attitude that core storage was a superior technique, and that he and his colleagues were being caught in the crossfire. When we tried, gently enough, to point out to him that core storage was simple, superior, and satisfactorily tested, it merely drove him closer to apoplexy.

It was August 1954 before IBM successfully tested its first magnetic core memory, which would be used in its XD-1 machine. Soon, by popular demand, the vacuum tube memories were being ripped out of machines across America to be replaced by a much smaller box of magnetic cores. The difference in speed and size was so noticeable that everyone wanted to have the new machines.

Big corporations did not buy computers in these early years of the industry; they leased them. So it fell to the likes of IBM and Bendix Systems to push through the upgrade to the new memory system. They did so, even though the patent litigation between MIT and RCA over who owned the design to this memory system rumbled on in the background.

Just about every computer in America was soon using magnetic core memories. It would be 1968, some fourteen years after IBM started dabbling with the technology, before the issues were finally resolved.

The dispute was not only about who invented the new memory first but also about all the modifications that had been made along the way, each of which had been patented individually. Most of those modifications were not designed by Forrester, but by Buck, Olsen, Papian, and the other researchers working on the project. The switch designed by Olsen that Buck had seen copied in Rajchman's lab was dragged into the dispute, as were other refinements to the design. Patents were filed and refiled to clean up the arguments. One of Buck's inventions was split into a separate patent filing, seemingly to clear up one aspect of the legal argument.

Meanwhile, the organization responsible for commercializing MIT's work remained doggedly focused on the enforcement of Forrester's memory patents for magnetic cores. Research Corporation, as the business was called, patiently awaited its payday from an expected settlement. The organization sent annual reports to MIT's inventors detailing the income and expenditure on their patents.

For years the income line read "none" while the expenses grew ever higher. The only commentary on the magnetic cores and the affiliated patents was that "no effort will be made to license his patent application until such time as the Patent Office indicates the scope of the claims that may be allowed."

With every year that passed, the potential payout grew bigger: by the early 1960s the technology was everywhere. IBM was worried; it was dominating the market with a disputed technology. It, too, had started to claim that it played a big role in developing the magnetic core memory technology in the first place in an attempt to chip away at the size of the bill it would inevitably have to pay to settle the dispute.

A crunch meeting between IBM and MIT on January 26, 1961, opened the door to a deal. Minutes of the showdown also reveal how the murky, fudged divide between the private and public sectors was still a problem.

James Birkenstock, one of the top advisers to IBM president Thomas J. Watson Jr., was the burgeoning computer giant's lead negotiator. He was up to his neck in IBM's work with the military. He mapped out the economics of the business for MIT's benefit; about 95 percent of IBM's computers were leased, he explained. The 5 percent who bought their own machine paid a price equivalent to fifty months' rental. Many customers purchased some parts, and leased the others.

Watson promised to send MIT a list of every machine that had been sold or leased, and to whom—including the US government. This was an important wrinkle in the talks: under the industrial cooperation agreement the government did not have to pay royalties on patents, although it did have to pay for the equipment.

"Mr. Birkenstock stated that it is sometimes difficult to know whether a particular machine is used by the government or a private organization," the minutes of the meeting note. "For example, a machine used by Rand Corporation may be at work on both Government and private contracts; or a machine used by Pan-American [Airlines] may be used in connection with missile testing at Cape Canaveral. Mr. Birkenstock indicated, incidentally, that IBM's charges are the same on Government and private machines."

The MIT negotiators pressed for details on what IBM thought it had done in the development of the magnetic core memory. Birkenstock pointed to IBM's role in the Whirlwind project and the military computer it subsequently developed to run the defense shield. He denied the MIT line suggesting that IBM had been trying to stifle the magnetic core technology. IBM had been working flat out on magnetic core memory, Birkenstock claimed, but "while all of MIT's developments were publicized, IBM's developments were kept secret, as is usually the practice in private industry."

Birkenstock went on to claim that MIT's technologies were only of interest to the military because he and his colleagues had pushed for the MIT work to be used rather than a rival technology developed by the University of Michigan.

"Mr. Birkenstock also insisted that computer development did not start at MIT, but that it was IBM who built the first computer, the SSEC [a successor to the Harvard Mark I], which was given to Harvard," the meeting minutes note. "The SSEC is a systems patent covering every computer in existence today."

Although Birkenstock claimed at the start of the meeting that he did not want to discuss a settlement, he soon started talking numbers. The MIT lawyers proposed that IBM pay two cents for every magnetic core it used.

According to the minutes of the meeting, "Mr. Birkenstock emphasized that IBM's operations in large-scale computers have been running at a loss since their inception. Later in the discussion, however, he said that a royalty of 2 cents a core would mean that some computers that are presently being sold at a profit would have to be sold at a loss."

Birkenstock explained that MIT had completely misunderstood the costs of IBM's computer business. Component costs were falling quickly, so the two-cent royalty would be equivalent to 4 percent of sales. The total cost of the machine was not just about parts, however. IBM sold electric typewriters at a markup of three times manufactured cost, and computers at ten times the manufactured cost, he explained. Yet a "mark-up of such magnitude is made necessary by the high cost of selling and of customer assistance."

A royalty payment equivalent to 4 percent of sales would have "serious consequences" for IBM, Birkenstock said. But it would be "even more drastic for other members of the industry, possibly forcing them out of business." He then proposed an alternative settlement, based on a series of complicated sums. It worked out at a flat fee of between $1.5 million and $1.6 million to cover all past and future infringements of MIT's patent.

It would take another seven years of negotiations to hammer out the final deal between IBM and MIT, yet the broad structure did not really change. The talks were held up for several years until the original patent dispute between MIT and RCA was resolved. Eventually the argument was settled thanks to a paper written in 1950 by Forrester—the same paper that Buck had sent to Joseph Eachus at Seesaw days after first arriving at MIT. Irrespective of who came up with the idea first, it was clear that Forrester had written it down before anyone else. As a result, IBM decided to settle with MIT and Rajchman lost out.

The $1.5 million that Birkenstock had initially proposed ballooned by 900 percent, to more than $13 million in the final settlement, according to legal papers released by MIT in the research for this book. The settlement forced every other computer manufacturer that had been using the technology to reach their own settlement. In total, MIT received more than $25 million from legal settlements over the patent, equivalent to about $176 million in today's money.

As the sole inventor listed on the magnetic core memory patents, Jay Forrester was suddenly a rich man. MIT had a system where the inventor got a 12 percent cut of the royalty payments. As a result Forrester got about $3 million over the years, or about $21 million in today's money.

Ken Olsen got what he described in a letter to the Buck family years later as a "nominal sum" for his work on the switch. That "nominal sum" still ran to tens of thousands of dollars, however. A letter sent to Forrester on February 26, 1965, from Paul Cusick, MIT's comptroller, details the first payments made to the inventors of the magnetic core memory patent. MIT had received a check for $2.8 million, so Forrester was being sent $323,373—described as 12 percent of what he was owed, "less $10,707 which was delivered to Kenneth H. Olsen."

The formula by which Olsen's share was calculated is not possible to understand from the documents in MIT archives; a side deal of some kind appears to have been negotiated. It equates to almost $80,000 in today's money for that first installment alone.

Olsen started to receive these monies at the time he was setting up Digital Equipment Corporation, which went on to become one of America's biggest manufacturers of office microcomputers. It employed 140,000 people at its peak, and was eventually bought by Compaq in 1998 for $9.6 billion—at the time, the biggest deal ever in the world of computers.

According to sources familiar with the settlement, Olsen also gained the right to use the magnetic core patent for free as part of his negotiations with MIT in regard to the legal settlement in recognition of the fact that he had done a great deal of the work on the technology.

Although Buck played a similar role in developing the patents, he received nothing for his work. He was named in the documents and his patents helped to swing the deal, but neither Buck nor his family received a penny from the settlement.

Internal correspondence between Forrester and the MIT team running the legal case suggests that Buck may have been blamed for sharing too much information about the secrets of Project Whirlwind. Forrester knew that Buck was still in regular contact with his old colleagues at Seesaw. As the case started to reach the crunch phase of negotiation, he seems to have looked to Buck as the possible source of a leak.

"It was his [Buck's] duty to report to them information that might be of interest to them which was happening at the Digital Computer Laboratory, and he also made extensive trips to other organizations and lectured on the computers and components indicating what we were doing at the Digital Computer Laboratory," Forrester wrote before the case had been settled. "Buck's name should be added to the list of people whose telephone calls and travel vouchers are being traced by the accounting office. We should try to reconstruct all the organizations he went to during 1950 and the first half of 1951. In particular, his reports to the National Security Agency may contain very valuable information."

By the time IBM and MIT put pen to paper on their agreement, Buck had been dead for nine years. Birkenstock, the IBM negotiator, slipped in an extra condition late in the discussions: IBM also wanted its hands on "the Buck patent."

The patent in question was not for the light gun, or any of the modifications to the magnetic core memory patent; it was the technology that had brought Dudley Buck international fame and recognition. IBM wanted the right to use the cryotron—and it was duly granted. At the time, IBM still believed the cryotron would be the future of the computer.

Project Lightning ended up being about a lot more than just cryotrons. It always suffered a stigma from the fact that it did not produce a single machine that everyone could point to when senior figures in government asked where all the millions of dollars had gone. Yet it has been credited in several papers with providing the impetus for new ideas.

It was Project Lightning that concentrated the industry's minds on the idea that quicker circuits led to better computers. That in turn led to Moore's law, the thesis posited by Intel cofounder Gordon Moore that microchips double in speed every two years. Although there is no physical or scientific basis for Moore's law, it is an observation that has held true since it was coined, with those advances in speed leading to the extremely mobile electronics of the twenty-first century.

Snyder's official history of NSA computing projects cites Project Lightning in glowing terms for this very reason. It mentions how the cryotron "proved not to scale to high speed operation as had been hoped." The detailed explanation of how the cryotron was used and what went wrong with it remains classified. It seems that it never was used as a missile guidance system, in spite of the time that was spent on the idea; the semiconductor took that crown.

Yet the cryotron retained a hard-core group of fans among the senior ranks of the American science establishment.

IBM spent years—and something of the order of $250 million—working on superconducting microchips. The cryotron evolved into a device called the Josephson junction by the mid-1960s. In 1987, scientists at IBM won international acclaim for creating a superconductor that could switch between states at much higher temperatures—opening the door to superconductors that could operate at room temperature. In February 2012 the company's Watson Research Lab unveiled a superconducting quantum computer. Although the theory behind it relies on the mad world of quantum mechanics, the core materials used in the chip are silicon, aluminum, and niobium—the same materials Buck had been experimenting with fifty-three years earlier. It runs on modified cryotrons.

Intriguingly, in the fallow years when American researchers paid little heed to superconducting microchip research, Russian researchers made significant strides. The technology became central to the development of the Soviet Union's most advanced computer chips through to the fall of the Berlin Wall in 1989.

As David Brock of the Computer History Museum explains,

> The Russians kept up with superconducting electronics. They were consistently into it in the same way that the NSA was into it. IBM tried to build this computer with the next generation of cryotrons, which were originally called tunneling cryotrons, and now they are called Josephson junctions. IBM had this gigantic project; that failed, not really technically, but IBM did not see it as commercially viable, so it died.
>
> Until the recent interest in quantum computing—much of which is based on these tunneling cryotrons—the technology had lain dormant in the US. But the Russians always kept going with it. In fact, a lot of the new stuff that's in superconducting electronics and the superconducting approach to quantum computing are from Russians who got out after the fall of the Berlin Wall. I have no doubt that the Soviets were really interested in cryotrons, because they kept with it.

In 1985, researchers at Moscow State University outlined a theory for a new superconducting chip it called rapid single flux quantum. It was a faster, more energy-efficient interpretation of the Josephson junction—that is, another modified cryotron.

By 1997 Moscow State University had formed a partnership with Bell Laboratories (with backing from the NSA) named the Hybrid Technology Multi-Threaded project, which was tasked with finding a replacement for silicon to produce petaflop-paced supercomputers. Four decades after Dudley Buck's death, the NSA still believed the cryotrons could beat a silicon transistor. That specific project ended in 2000.

By the early 2010s, however, NSA researcher Marc Manheimer persuaded the agency to have another look at superconducting super-computers. Manheimer, based at the NSA's Laboratory for Physical Sciences, almost immediately next door to NSA headquarters in Maryland, has said that he encountered multiple skeptics due to a "history of failure" with Dudley Buck–inspired technology.

Yet by 2013 Manheimer had won over Intelligence Advanced Research Projects Activity (IARPA), the research department for the intelligence community, to create the Cryogenic Computing Complexity program. He switched agencies to run the program, the budget for which has never been disclosed.

Public disclosures from 2014 show that defense titans Northrop Grumman and Raytheon were awarded a slice of the contract, along with IBM. And much of the work is being conducted in MIT's Lincoln Laboratory—where Buck began work on Project Whirlwind in 1950.

The NSA and IARPA are not the only agencies to have continued dabbling with Buck's technology; superconductors remained popular with NASA for years after his death. Wernher von Braun, the German-born mastermind of the V-2 flying bomb who had switched sides after World War II to lead missile and space research in the United States, had something of an obsession with the subject.

The Josephson junction—the successor to the cryotron—was used by NASA for years. Although it was a different invention, Braun insisted on still calling the new switch a cryotron. NASA technical papers suggest that the space agency continued research into these cryotron-like switches until at least the mid-1990s.

Braun wrote in the January 1969 edition of *Popular Science* magazine about how magnets and superconductors could be used to create a force field that would protect spaceships from solar flares on future missions to Mars. Braun made it quite clear how he thought these future spaceships would find their way:

> The propensity of many superconducting materials to "go normal" (lose their superconductivity) in a magnetic field has one

useful and redeeming aspect. It permits their employment as con-
tactless switching devices, called cryotrons. A cryotron consists
of a thin-film "gate wire" and a "control wire," both superconductive. Send a current through the control wire, and its magnetic
field kills the superconductivity of the gate wire, giving the effect
of an on-off switch.

All basic types of electronic computers' circuits can be built
from combinations of these microminiature switching units. The
resulting computer, which is kept refrigerated in operation, is re-
duced to shoebox size and consumes extraordinarily little
power—ideal qualities for space use. A cryotron computer and a
superconducting gyro and accelerometer could make up a high-
precision navigation system to help future astronauts find their
way about the solar system.

At a recent conference in Abu Dhabi, a former administrator of
NASA was asked over dinner about cryotrons and Josephson junc-
tions. He hesitated, scowled, then barked: "How the hell do you know
about that?"

21
POSTSCRIPT

J ACKIE BUCK HAD TO GET BACK ON HER FEET QUICKLY. THERE HAD been a little bit of money saved up, but not much. None of Buck's patents were producing any income of note, although the MIT patent office would keep her up to speed on its progress nonetheless.

The dean of one of the local graduate schools, a friend of the family, wanted to help out when he learned of Buck's death. He let Jackie study social services on a part-time basis—it was the first time the school had ever let anyone study part-time.

Spreading her study over a few years, she eventually got her degree and started working for the child guidance movement, then for Massachusetts General Hospital before applying for a job in the guidance department at MIT. She spent thirty years working in the same buildings where her late husband had performed his experiments.

Jackie started to call her neighbors in Wilmington her angels. When the babysitter couldn't make it or she got held up at work on short notice, it was Nancy Bodenstein, Jeanne Collins, and Shirley Harding—the same three women who had gone with her to see Fidel Castro speak—who would lend a hand.

Jeanne Collins had seven children of her own, but would quite often make space for the Bucks. As Jackie remembers, "Jeanne, who cooked every day for nine, thought nothing of adding another four to her table, and wouldn't take no for an answer. Likewise, Nancy Bodenstein would insist on sharing their evening meal. Nancy would apologize, saying that it was a 'pretty simple meal.' I told her that it tasted like ambrosia. After a long day and commute, it was such a help not to have to prepare a meal, in addition to the nightly baths, putting-to-bed ritual, laundry, and preparing generally for the next morning."

With Jackie working at MIT, many of Buck's old friends stayed in close touch with the family. When the family outgrew the house in Wilmington, they rented a bigger place at beneficial rates from Harold "Doc" Edgerton, the MIT professor who pioneered the strobe light and worked on subsea cameras with Jacques Cousteau.

Carolyn, Douglas, and David only knew their father through other people's stories. Every day after school they would wander the corridors of MIT, waiting for their mother to finish work. Douglas in particular liked to look for his dad's old equipment.

"I remember once when he was leading around the president of MIT, James Killian [Eisenhower's former science adviser], and they were looking for a cryotron probe that Doug was sure was there in a closet somewhere," Jackie says. "Everyone was so fond of the children, because they were so fond of Dudley. Killian could see that Doug was earnest in trying to find his dad's things."

Jerome Wiesner, the dean of science and a trusted adviser to President John F. Kennedy, also knew the Buck kids by name and would always say hello.

If Jackie was struggling to find her children, she usually tracked them down in Doc Edgerton's lab. When Edgerton wasn't building equipment for Jacques Cousteau, he would let them play around with his cameras. Doug, Carolyn, and David would take turns photographing drops of milk falling with Edgerton's rapid-fire, freeze-frame cameras. They would line up shots for his famous experiments photographing bullets passing through apples, bananas, and playing cards.

Doug was only two years old when his father suddenly died. Growing up, he was always hearing remarks from his father's former colleagues about the legacy he had left behind. Yet he had little idea of the specifics.

In 1975 Doug Buck decided to start gathering together more information about his father's life with the aid of another former MIT professor, Bernie Widrow. The Internet was starting to change the world, and the roll call of pioneers from the 1950s and 1960s did not include Dudley Buck.

Doug wanted to change that. Piece by piece, he procured Dudley's lab books and notes from MIT. With Bernie Widrow's help, Doug tracked down other people who had known his father. He sifted through boxes of files in his mother's attic, finding his father's diaries and correspondence. He also found a large collection of travel receipts recording journeys all over America—including frequent trips to Washington, DC. It was there too that Doug found the cryptic military papers that later transpired to be orders for two covert missions conducted in the Eastern bloc for an unofficial wing of the CIA.

Not long after Doug began his research, Killian approached him at an MIT function. He punched Doug on the shoulder and said: "Your father invented everything."

LIST OF INTERVIEWEES

Alport, Gerald W.
Aronson, Joel B.
Baghdady, Dr. Elie
Balmer, Don
Bannecker, Bill
Barneich, Ed
Bensky, Lowell
Berggren, Prof. Karl
Bremer, John
Brock, David
Brown, David R.
Buck (Harrison), Carolyn
Buck, Jacqueline
Buck (Schick), Virginia
Campbell, Glenn
Clark, Wesley A.
Collins, Dean
Comick, Tom
Cooke, Betsy (Helen Elizabeth Connelly)
Crawford, Prof. Charles K.
Eachus, Barbara
Eachus, Joseph
Endahl, Chuck
Fix, Bill
Gerovitch, Slava
Goodenough, John B.
Herzfeld, Fred

Honn, Rich
Huff, Lynn
Huskey, Harry
Jenney, Richard F.
Khrushchev, Sergei
Lee, Bob
Lowthian, Kenneth
Manheimer, Dr. Marc
Maxfield, Rev. Otis
Meadows, Lee
Menyuk, Norman
Nahman, Dr. Norris
Nelson, Norman
Olsen, Ken
Pacela, Dr. Allan F.
Papian, William N. (Bill)
Petersen, Dean
Petersen, Doris
Sass, Andrew R.
Schick, Georg
Schick, John
Schneider, Arthur John
Shoulders, Ken
Simpson, James
Stiening, Rae F.
Swain (nee Schupbach), Carol
Teiger, Shushan
Temme, Don
Tetrault, Genevieve
Whitman, Howard
Whitman, Nancy
Widrow, Dr. Bernard
Wigington, Ronald Lee
Zaorski, Ralph W.

CHAPTER NOTES

1: PROJECT LIGHTNING
The narrative of the Russian scientists' trip is drawn from an article entitled "Russian visit to US computers" in *ACM Magazine*, Volume 2, Issue 11, Nov. 1959. Link here: http://dl.acm.org/citation.cfm ?id=368498.

General background on Soviet Russia's attitude to computers and that of Josef Stalin was drawn from web research, CIA archives, and aided by passages in *From Newspeak to Cyberspeak, a History of Soviet Cybernetics,* by Slava Gerovitch, MIT Press, 2002. Professor Gerovitch was also interviewed by email and telephone.

2: SANTA BARBARA SOUND SYSTEM
The activities of the Santa Barbara Sound System were described in detail by Lee Meadows, who kept copies of the relevant newspaper clippings. The key events were also recalled by multiple family members.

3: THE V-12 PROGRAM
The V-12 program is well documented. The specific quote attributed to Vice Admiral Randall Jacobs came from "Navy V-12, Vol. 12." (Turner Publishing Co., 1996, Henry C. Herge). As is evident from the text, several members of the program who studied with Dudley Buck provided some additional anecdotes and detail regarding their experiences.

4: SEESAW
The account of the incident in the truck in Washington is as told by Glenn Campbell, Dudley Buck's foster son. Campbell frequently

went on unofficial ride-alongs with Buck. Although he was not there on this particular day, he was told of the story by the other officer as well as Buck.

The history of the Mount Vernon Seminary for Women is documented on the George Washington University website: www.library .gwu.edu.

The history of Operation Venona is well documented in *The Venona Story*, by Robert L. Benson, a web publication for the Center for Cryptologic History based on declassified NSA files.

The Neglected Giant: Agnes Meyer Driscoll, by Kevin Wade Johnson, a paper for the Center for Cryptologic History, details the incredible history of the most influential woman codebreaker of her day.

The biography of Howard Campaigne is sourced from an NSA oral history interview, NSA-OH-14-83, https://www.nsa.gov/news-features/declassified-documents/oral-history-interviews/assets/files /nsa-oh-14-83-campaigne.pdf.

The biography of Solomon Kullback comes from NSA oral history interview NSA OH-17-82 with Solomon Kullback, conducted by R. D. Farley and H. F. Schorreck on August 26, 1982, declassified December 9, 2008.

The biography of Joseph Eachus is taken mostly from an obituary in *The Daily Telegraph*, London, December 19, 2003.

Background information on the ENIAC and EDVAC machines and the formation of ERA is taken mostly from *History of NSA General-Purpose Electronic Digital Computers* by Samuel S. Snyder, 1964, https://www.nsa.gov/news-features/declassified-documents/nsa -early-computer-history/assets/files/6586784-history-of-nsa-general -purpose-electronic-digital-computers.pdf.

5: OPERATION RUSTY

Reinhard Gehlen's story is well documented. The CIA's online archives on Gehlen are extensive. In particular, the declassified account by the officer who orchestrated his switching of sides in 1945: "Report of Initial Contacts with General Gehlen's Organisation," by John R. Boker, 1 May 1952, Part I.

While there are many available sources of information on Konrad

Zuse, the most comprehensive is the Konrad Zuse Internet Archive, found at www.zuse.zib.de.

Zuse spoke about America's attempts to secure his services in an oral history interview with the IEEE Center for the History of Electrical Engineering, conducted by Frederik Nebeker, August 28, 1994.

6: PROJECT WHIRLWIND
Jay Forrester's paper, "Forecast for Military Systems using Electronic Digital Computers," was declassified on June 8, 1956. It is available on archive.org.

MIT has produced an excellent online archive regarding Project Whirlwind, available at the following link: http://museum.mit.edu /150/21.

The account of the cryptological chaos surrounding the Korean War is taken from *American Cryptology during the Cold War; 1945-1989, Book I: The Struggle for Centralization 1945-1960*, Thomas R. Johnson, Center for Cryptologic History, NSA, 1995. The specific quote used is found on p.40 of the document, which can be found in the NSA's online archives—and remains heavily redacted.

Buck's work on the light gun is recorded in his lab books. External verification of his role in its development was found in the *Encyclopedia of Library and Information Science* Volume 5 by Allen Kent and Harold Lancour, p.396. It reads: "A historical footnote, incidentally, was that Dr. Dudley Buck was the pioneer most responsible for the light gun—the same Dudley Buck who later made such great contributions to micro-miniaturized circuitry and whose death at a young age terminated a career just beginning to blossom."

Video footage of the Whirlwind computer's CBS interview with Edward Murrow on December 16, 1951 can be found on YouTube.

7: MEMORY
The story of Glenn Campbell came almost entirely from Glenn himself. While modern readers may feel inclined to be suspicious of the relationship between a young bachelor and a teenage boy, any such doubts are wholly unfounded. As described by Glenn, Dudley saved him from a life of neglect and abuse. He remained

extremely grateful for that in his advanced years.

General background on the capabilities of early computers is taken mostly from *History of NSA General-Purpose Electronic Digital Computers* by Samuel S. Snyder, 1964, https://www.nsa.gov/news -features/declassified-documents/nsa-early-computer-history/assets /files/6586784-history-of-nsa-general-purpose-electronic-digital -computers.pdf.

The correspondence between James Forrester and Captain J. B. Pearson, deputy director of the Office of Naval Research, was among a bundle of documents released from the MIT archives pertaining to a legal dispute over the magnetic core technology, a theme that is expanded upon later in the book.

8: FORGING BONDS

Biographies of the attendees at the summit in Corona, California, are comprised from an amalgam of Buck's briefing notes prior to the event and contemporaneous web research.

Similarly, the properties of "heavy water" and deuterium have been outlined based partly on descriptions in Buck's correspondence— backed up with web research.

9: PROJECT NOMAD

The impact of the Korean War on America's security services is widely described in multiple declassified files in the CIA archives, and has been commented upon before.

The specific agenda of Yehoshua Bar-Hillel's 1952 machine translation conference has been posted online by MIT archivists—including papers that were presented at the event, and materials circulated beforehand: http://mt-archive.info/MIT-1952-TOC.htm.

Howard Campaigne's NSA oral history interview can be found here: https://www.nsa.gov/news-features/declassified-documents/oral -history-interviews/assets/files/nsa-oh-14-83-campaigne.pdf.

Other background on Project Nomad is found here: https://www .nsa.gov/news-features/declassified-documents/crypto-almanac -50th/assets/files/NSA_Before_Super_Computers.pdf.

The specific figure regarding the $3.25m bill for the project is found on p. 29, *History of NSA General-Purpose Electronic Digital Computers* by Samuel S. Snyder, 1964.

This book does not attempt to offer a comprehensive history on every computer project in the US during the late 1940s and early 1950s. Snyder, however, did exactly that. Anyone interested in learning more should read his account, which runs to little more than 100 pages.

The discussions regarding Dudley Buck's contact with Amtorg and his request to travel to Moscow are discoverable in the CIA's historical archives by searching for "Buck," and scrolling through several pages. The following links should work: https://www.cia.gov /library/readingroom/docs/CIA-RDP79-01041A000100020160 -6.pdf; https://www.cia.gov/library/readingroom/docs/CIA-RDP79S0 1057 A000500070019-1.pdf.

10: TWO SMALL WIRES

The description of the discovery of superconductors and their properties was detailed in Buck's lab notebooks and have been fact-checked against contemporary sources.

11: THE CRYOTRON

David Brock has written an article for *IEEE Spectrum* magazine previously detailing the NSA's attempts to build superconducting computers, and establishing the chain of discoveries that leads from Dudley Buck to the present day: https://spectrum.ieee.org/tech-history /silicon-revolution/will-the-nsa-finally-build-its-superconducting -spy-computer.

The following slides prepared by MIT also chart the evolution of the cryotron: https://ocw.mit.edu/courses/electrical-engineering-and -computer-science/6-763-applied-superconductivity-fall-2005/lecture -notes/lecture15.pdf.

12: LAB RATS

The original 1956 press release from MIT about the creation of the cryotron was found in Dudley's files. The university frequently lists Prof. Dudley Allen Buck in articles about its greatest alumni.

The December 1988 edition of the university's *RLE Currents* newsletter provides a summary of Buck's work and refers to some of the papers that have been used in the research of this book: http://www .rle.mit.edu/media/currents/2-1.pdf.

13: THE MISSILE GAP

The description of the incident at Kapustin Yar comes from pp. 280-1 *Khrushchev: The Man and His Era,* William Taubman.

The portrayal of Von Braun's complaints about isolation is consistent across multiple accounts. The specific point about Hamill instructing Von Braun to avoid mingling with Americans comes from *Wernher Von Braun: Rocket Visionary*, revised edition 2008, Ray Spangenburg, Diane Kit Moser, Chelsea House Publishers, p.83.

A declassified account of the years-long efforts by the US and UK to gain access to Soviet intelligence and decrypt it is discoverable via the Internet Archive in a document entitled "Joseph McCarthy and the Venona Documents," authored by Cecil Phillips and Lou Benson. The specific quotes used are taken from this document, although it is not formatted in a manner that facilitates page numbers: https:/ /archive.org/stream/VenonaDOCUMENTSPARTS12/Venona%20 -%20DOCUMENTS%20-%20PARTS%201%20_%202-_djvu.txt.

The description of Louis Ridenour's life and achievements is taken mostly from an article by Dr. W. F. Whitmore in the Spring 1981 edition of *Horizons Magazine*, an internal Lockheed publication. The achievements of the SCR-584 radar system are well documented. A web page named "The SCR-584 Radar Tribute Page" collating useful sources of information about the device, its design, and deployment, can be found here: https://web.archive.org/web/20101004010816/ http://www.hamhud.net/darts/scr584.html.

Wernher von Braun's vision of satellites carrying nuclear weapons that could be sent down to earth is well established. The particular quote about the pressing need to reach space first has been widely used by many of his biographers, including Michael Neufeld in *Von Braun: Dreamer of Space, Engineer of War,* Knopf Doubleday Publishing Group, 2017, pp. 221-222.

14: FAME

A file of press cuttings about Dudley's achievements was stored with his personal effects in Jackie Buck's attic. The *Wall Street Journal* article quoted was also in Dudley's files, even though he was not referenced.

Project Vanguard is explained extensively in the history section of NASA's website, and appears to be accurately summarized by Wikipedia. References to Dr. Maurice Dubin's experiment can be found here, for example: https://history.nasa.gov/SP-4215/ch1-3.html.

IBM's work in superconducting chips is partly detailed in Snyder's history of NSA computer projects. IBM historians have also documented the company's involvement in superconducting computers, which is where the accounts of work by Crowe and Garwin is taken from. The particular statistics about switching speeds came from *Superconductivity at IBM - a Centennial review: Part I - Superconducting Computer and Device Applications*, William Gallagher, p. 3.

15: THE POST-SPUTNIK EFFECT

The account of the Sputnik reaction at a US cocktail reception is taken from NASA historian Roger Launius in his paper "Sputnik and the Origins of the Space Age," available on the history section of the NASA website: https://history.nasa.gov/sputnik/sputorig.html. The quote attributed to George Reedy, and the general background to the creation of NASA, is also taken from Launius's paper.

The background regarding the visit of the Russian computer scientists was laid out in the article referred to earlier, "Russian visit to US computers" in *ACM Magazine*, Volume 2, Issue 11, Nov. 1959.

16: A RECIPROCAL ARRANGEMENT

The account of the arrangements made for the Soviet scientists to come to the US, and their attempts to attend the Eastern Joint Computer Conference, is drawn from "Russian visit to US computers" in *ACM Magazine*, Volume 2, Issue 11, Nov. 1959.

17: THE MISSILE MEN

Details of the WS117L program and Project Corona have been declassified and extensively reported. The specific analysis of how the project was divided into three was taken partly from "A sheep in wolf's clothing: the Samos E-5 recoverable satellite (part 1)" by Dwayne Day, *The Space Review*, July 6, 2009.

18: THE RUSSIANS HAVE LANDED

The account of the Russian scientists' trip is again drawn from an article entitled "Russian visit to US computers" in *ACM Magazine*, Volume 2, Issue 11, Nov. 1959.

19: A PACKAGE

Fidel Castro's 1959 trip to the US was a highly public event. Everything mentioned here is taken from contemporaneous news reports.

20: THE EXTRA PIECES OF THE PUZZLE

Howard Aiken's quotes are taken from an interview for the Smithsonian's Lemelson Center for the Study of Invention and Innovation, National Museum of American History, Computer Oral History Collection, 1969-1973, 1977, Howard Aiken interview conducted by Henry Tropp and I. B. Cohen, February 1973.

The CIA paper for the Kennedy commission is entitled "Soviet Use of Assassination and Kidnapping" and was declassified on September 22, 1993. It can be found in the archives section on the CIA website, currently at this address: https://www.cia.gov/library/center-for-the-study-of-intelligence/kent-csi/vol19no3/html /v19i3a01p _0001.htm.

Descriptions of Shelepin's rogue operations can be found in *Khrushchev: the Man and His Era*, by William Taubman, pp. 468-470.

The account of the Magnetic Core legal dispute is taken from legal correspondence released by the MIT archives, with additional background information from interviews with individuals who were involved in the case.

The Jan Rajchman interview cited was conducted on October 26, 1970 by Dr. Richard R. Mertz as part of a series for the Smithsonian's Lemelson Center for the Study of Invention and Innovation, National Museum of American History, Computer Oral History Collection, 1969-1973, 1977: http://amhistory.si.edu/archives/AC0196_rajc701026.pdf.

ACKNOWLEDGMENTS

THE COMPLETION OF THIS BOOK WOULD NOT HAVE BEEN POSSIBLE without the assistance of so many people whose names cannot all be enumerated. Their contributions are sincerely appreciated.

All of my relatives were encouraging and supportive. There are so many others to thank. Many are dead.

In particular, I would like to extend thanks to Bernie Widrow, Don Burrer, Alan Pacella, Chuck Crawford, Harold Edgerton, Lee Meadows, Tom Comick, Professor Karl Berggren, and to Howard and Nancy Whitman.

I must thank Alan Dewey, and all the people who agreed to be interviewed

I would also like to thank Iain Dey, who raised his hand at the Palm Beach Grill one night.

For all the family members, friends and others who in one way or other shared their stories about my father, thank you.

—DOUGLAS BUCK

INDEX

Bell Laboratories (Bell Telephone Laboratories): and Buck, 87, 95; and Moscow State University, 255; and transistor, 77
Bendix Systems, 206, 215, 249
Bensky, Lowell, 247–48
Benson, Lou, 145
Berggren, Karl, 175–76
Berlin, 41–42, 51
Bers, Lipman, 207, 211, 217
BESM, 208
BESM II, 208
binary code, 14, 76
Birkenstock, James, 250–53
Bishop, L. G., 226
Bismutron, 107–9
Bissell, Richard, 203
bit, 76
Black, David, 201
Bletchley Park, 45–46; and bits, 76–77; and Turing, 98
Blois, M. Scott, 92–94
Bodenstein, Nancy, 83, 222, 233, 259
Boeing, 15, 158, 226–27
Boker, John R., 53
"Bold Strategy to Beat Shortage, A" (Killian), 136
Bolshaya Elektronnaia Schetnaia Mashina (BESM), 208
Bolster (admiral), 68
Boolean logic, 76
Booth, Andrew Donald, 97
Braun, Wernher von: and Explorer 1, 169; and Popu-
lar Science, 256; and prisoners of peace, 143–44; and satellite missiles, 152, 162; and superconductors, 256
Brock, David: on Buck and ABEL, 48; on Buck, death of, 234–35; on Buck and mathematical congress in Darmstadt, 55; on Buck and superconducting, 121; on Buck as pioneer, 89; on Buck, reputation of, 175; on content-addressable memory, 100–1; on the cryotron, 170–71; on NSA and computers, 88; on Project Lightning, 174; on Russian chips, 255; on transistor, 99
Brown, David, 102
Brown, Gordon, 155–56, 185, 195, 231
Buck, Allen, 21, 194
Buck, Carolyn, 122, 194
Buck, David, 202, 220, 228, 231, 260
Buck, Delia, 22–26, 90, 114, 130, 221
Buck, Douglas, 156–57, 194, 260–61
Buck, Dudley H. (uncle), 220
Buck, Dudley: and ABEL, 48; and Air Force Project 438L, 195; and Amtorg, 214; and Arthur D. Little, 117–18; background of, 16; birth of, 21; and Bell, 87; and